Designing with Photovoltaics

Designing with Photovoltaics

Edited by
Angèle Reinders

CRC Press
Taylor & Francis Group
Boca Raton London New York

CRC Press is an imprint of the
Taylor & Francis Group, an **informa** business

CRC Press
Taylor & Francis Group
52 Vanderbilt Avenue,
New York, NY 10017

and by

CRC Press
2 Park Square, Milton Park, Abingdon, Oxon OX14 4RN

Printed on acid-free paper

International Standard Book Number-13: 978-1-138-29973-3 (Hardback)

Visit the Taylor & Francis Web site at
http://www.taylorandfrancis.com

and the CRC Press Web site at
http://www.crcpress.com

Contents

Preface

A smooth energy transition from fossil fuels to renewable energy sources is one of the principal challenges that mankind faces at the moment. It will delay and, hopefully, stop climate change caused by the emission of large quantities of CO_2 and other greenhouse gases. Namely, these emissions largely originate from our fossil energy demand, which as such should be drastically reduced. In 2015, it has therefore been agreed upon by the Paris Agreement (UN 2015) that the global temperature rise should stay below 2°C, actually preferably below 1.5°C, compared to preindustrial levels. As a consequence, the required changes from a fossil-fuel society to one based on renewable energy will involve a huge effort by a varied group of stakeholders: the public, policy makers, scientists, industry, engineers, and designers. Many international and national agreements state that solar photovoltaic (PV) energy technologies will be one of the main contributors to achieving a prospective 100% renewable energy supply. And it is generally believed that both societal acceptance and technology development will be key to realizing this goal. Also, it is known that the interdisciplinary field of design can bring these two aspects together, by the creation of products or systems, which people like to use (Reinders et al. 2012). As such, we envision that this book entitled *Designing with Photovoltaics* will inspire its readers to more frequently and better apply a wonderful, sustainable solar-energy technology, such as photovoltaic (PV) cells, in a wide range of existing and future things and objects, better said, products. Namely, the aim of this book is to inform the reader such that (s)he will be able to easily understand how new products, buildings, and vehicles containing PV technologies can be designed. Hopefully in the end, the reader would like to design new PV-powered products by her/himself. Therefore, this book covers a broad range of topics related to the design of products, buildings, and vehicles with integrated PV technologies.

Two clear examples of how PV-powered products are used in our society today are (1) PV façade in buildings and (2) solar charging of electric vehicles. In both cases, the emissions of greenhouse gasses due to energy consumption of respectively buildings and transportation can be minimized by the application of solar cells, which yield very low, close-to-zero emissions of CO_2 during electricity production over their full lifetime. In these two cases, and many others as well, PV technologies can be categorized as "cleantech." Therefore, to reduce greenhouse gas emissions and hence endeavor to retard climate change, inevitably we have to learn how *photovoltaic technologies* can be better applied in products, buildings, and vehicles.

This book is a bit different from previous publications in the field of PV research and design engineering, namely:

- Semiconductor physics of PV cells and their underlying materials won't be a major topic of this book. For these items, we refer to *Photovoltaic Solar Energy—From Fundamentals to Applications* (Reinders et al. 2017).
- This book won't be monodisciplinary; instead, it will cover a broad range of interrelated topics that are relevant for the design of PV-powered things.

- This book is not meant to be a DIY handicraft book. The authors are all professionals and would like to share their experiences with real design processes, design research, and the creation of real products, vehicles, buildings, and other objects with its readers.
- This book won't focus on building-integrated PV only.
- This book won't focus on indoor PV only.

As a consequence, the authors believe that *Designing with Photovoltaics* has a unique approach and content, covering many different product categories with a focus on product design, which so far hasn't been written or published elsewhere.

For this book project, a team of excellent experts in the field of PV design and PV applications has been brought together, paying attention to specific design aspects by multiple design cases and a myriad of colorful, visual information, while at the same time also providing knowledge at an academic level.

Therefore, the content will be interdisciplinary, aiming at reaching a broad audience of skilled engineers, designers, and researchers. Most of all, we hope to inspire anybody interested in solar power and to be a rich comprehensive source of information about potential applications of PV technologies.

Please enjoy reading and please feel welcome to start designing our solar future and as such become part of the sustainable energy transition!

With best regards on behalf of all the authors,

Angèle Reinders
Eindhoven University of Technology
and University of Twente
The Netherlands

REFERENCES

Reinders, A.H.M.E., J.C. Diehl, and J.C. Brezet. (2012). *The Power of Design*, Wiley: West-Sussex. p. 331, hard-cover book, John Wiley & Sons: London.

Reinders, A.H.M.E., P.J. Verlinden, W.G.J.H.M. Van Sark, and A. Freundlich. (2017). *Photovoltaic Solar Energy: From Fundamentals to Applications*, hard-cover book, John Wiley & Sons: London.

United Nations. (2015). Paris Agreement. https://unfccc.int/sites/default/files/english_paris_agreement.pdf

Acknowledgments

We acknowledge all contributions so far by students, colleagues, and companies to the still-emerging field of designing of photovoltaics. In particular, Georgia Apostolou, Alessandra Scognamiglio, Wilfried van Sark, Marjan van Aubel, Cihan Gerçek, Wouter Eggink, and Ned Ekins-Daukes are warm-heartedly thanked for their enthusiastic, continued and persistent involvement during the realization phase of this book by writing chapters and design cases, as well as supporting the review of the main texts. Thank you so much; without your contributions, time, and energy, this book would never have seen the daylight.

Solar Team Twente, represented by Merel Oldenburg and Michael ten Den, thank you for sharing your insights in the development of solar racing cars in a special design case, which reveals the "secrets" of good solar car design. You were able to make time for your contribution, while simultaneously developing and producing a new car, which evidences your passion for applications of photovoltaics.

Also, Peter Krige Eelke Bontekoe, Joost van Leeuwen, Hanbo Yang, Monika Michalska, Rosina Pelosi, Marcello Nitti, Lara Gillan, Dane McCamey, Elham M. Gholizadeh, Parisa Hosseinabadi, Blair Welsh, and Scott Kable are thanked for their very important contributions to respectively. Design Case 1 about the Current Table, Chapter 5, and Design Case 2 on Luminescent Solar Concentrator PV designs, which was inspired by the visionary outlooks on exciton science by Tim Schmidt; Tim, thank you very much.

All companies and organizations who provided permission to use their visual materials in this book are gratefully thanked, in particular NeaStudio, Sono Motors, C.F. Møller Architects, and NASA. Last but not least, I would like to thank Jeroen Verhoeven, of DeMakersVan, who, with his incredible creativity, is able to transform a merely technical material such as silicon into seemingly living matter by his design of Virtue of Blue, which is presented in Design Case 3. Thank you for your inspiration.

Editor

Angèle Reinders is a full professor in design of sustainable energy systems at Eindhoven University of Technology in the Netherlands and an associate professor at the University of Twente, the Netherlands, where she runs a research group on sustainable energy and design. Her research focuses on design, product development, and system integration in the framework of sustainability and renewable energy technologies. In particular, approaches are explored that support a better integration of new energy technologies in products, buildings, and local infrastructure with the purpose of increasing sustainable energy use and energy-efficient behavior by end users in the context of living, working, and mobility. She has practical experience with applications of photovoltaic (PV) solar technologies, fuel cells, energy storage technologies, LEDs, etc. In her ongoing projects, she focuses on smart grids and PV applications, such as PV systems, PV modules, PV-powered boats, building-integrated PV, and product-integrated PV. She has published more than 100 papers, edited two books, and is a cofounding editor of the *IEEE Journal of Photovoltaics*. She was the technical program chair of IEEE-PVSC-40 in 2014 and the chair of the IEEE-PVSC-44 in 2017. She has a vast international experience and stayed at Fraunhofer ISE (Germany), World Bank (US), ENEA (Italy), Jakarta and Papua (Indonesia), and the Centre for Urban Energy (Canada) for her research. She holds a master's degree in physics and completed a PhD in chemistry from Utrecht University, the Netherlands. At present, she teaches in the Study of Industrial Design Engineering and the Master of Sustainable Energy Technology. With students, she explores the possibilities of designing with photovoltaics in various design and research projects.

Contributors

Georgia Apostolou
Aristotle University of Thessaloniki
Thessaloniki, Greece
and
University of Twente
Enschede, the Netherlands

Eelke Bontekoe
Utrecht University
Utrecht, the Netherlands

Wouter Eggink
University of Twente
Enschede, the Netherlands

Ned Ekins-Daukes
University of New South Wales
Sydney, New South Wales, Australia

Cihan Gerçek
University of Twente
Enschede, the Netherlands

Elham M. Gholizadeh
University of New South Wales
Sydney, New South Wales, Australia

Lara Gillan
University of New South Wales
Sydney, New South Wales, Australia

Parisa Hosseinabadi
University of New South Wales
Sydney, New South Wales, Australia

Scott Kable
University of New South Wales
Sydney, New South Wales, Australia

Peter Krige
Imperial College London
Royal College of Art
London, England

Dane McCamey
University of New South Wales
Sydney, New South Wales, Australia

Monika Michalska
Monash University
Melbourne, Victoria, Australia

Marcello Nitti
Eindhoven University of Technology
Eindhoven, the Netherlands

Merel Oldenburg
University of Twente
Enschede, the Netherlands

Rosina Pelosi
University of New South Wales
Sydney, New South Wales, Australia

Timothy Schmidt
University of New South Wales
Sydney, New South Wales, Australia

Alessandra Scognamiglio
ENEA
Naples, Italy

Michael ten Den
University of Twente
Enschede, the Netherlands

Marjan van Aubel
Marjan van Aubel Studio
Amsterdam, the Netherlands

Joost van Leeuwen
Utrecht University
Utrecht, the Netherlands

Wilfried van Sark
Utrecht University
Utrecht, the Netherlands

Jeroen Verhoeven
Dutch Design House DeMakersVan
Amsterdam, the Netherlands

Blair Welsh
University of New South Wales
Sydney, New South Wales, Australia

Hanbo Yang
University of Melbourne
Melbourne, Victoria, Australia

1 Introduction

Angèle Reinders, Wouter Eggink, and Marjan van Aubel

CONTENTS

1.1 A SOLAR REVOLUTION

The idea of integrating solar technology into objects, such as products, buildings, or vehicles (Figure 1.1) is not new. Already in the 1950s people thought about integrating solar photovoltaic (PV) cells into smaller products that were not connected to the household's mains to let them autonomously generate electricity for these products. Unfortunately, this never happened because electricity from the grid was cheap, and the price of germanium, a basic element in solar cells at that time, was too high. Therefore, the focus of embedding PV technology into small household objects shifted to integrating them into space applications such as satellites. Here, having solar panels onboard made much more sense because it meant that some functions of the satellite could function independently using the power of the sun. In this context, photovoltaic technologies result in the generation of power from photons, the smallest energy packages contained by solar irradiance. In space, higher risks and costs were involved compared to the everyday household objects allowing the use of expensive advanced PV materials, such as gallium arsenide, in solar cells applied in satellites. PV developments in space have been dazzling, for instance, the International Space Station contains more than an acre of solar arrays powering the station, making it the second brightest object in the night sky after the moon, according to NASA (Figure 1.1). Since the development and production of solar technology, now nearly 70 years ago, the focus has been on price reduction and increasing the efficiency of the solar PV technologies. Scale has been an important factor. Something that is new, and only available in a small volume, is expensive, whilst by producing on a mass scale, prices will lower. In the last 40 years, each time the cumulative production of the silicon PV modules doubled, the price went down by 24% (FhG-ISE 2019), see Figure 1.2.

FIGURE 1.1 From left top clockwise shown: (a) the PV-powered International Space Station (Courtesy of NASA), (b) PV-powered building, named Copenhagen International School (Copyright by C.H. Møller Architects), and (c) the Sion, a PV-powered EV (Copyright by Sono Motors).

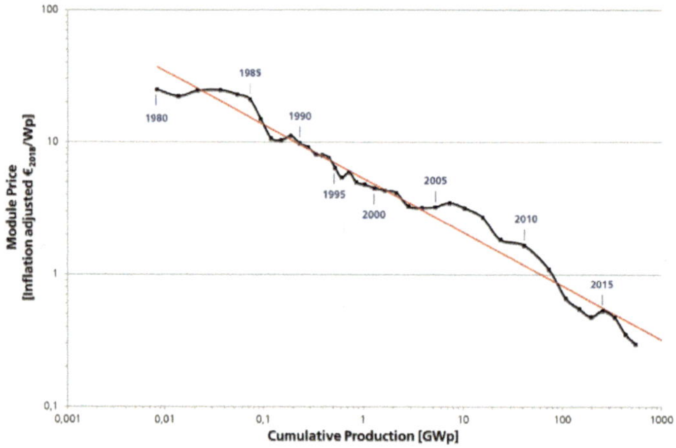

FIGURE 1.2 Learning curve of PV modules of all PV technologies showing PV module price versus cumulative production (FhG-ISE 2019). Data from 1980 to 2010 estimation from different sources: Strategies Unlimited, Navigant Consulting, EUPD, pvXchange; from 2011: IHS. Graph: PSE GmbH 2019.

Now that solar PV modules are being produced on a very large scale, their price has decreased immensely. To give an idea, at present (in 2019), a nominal power of 600 GW of PV systems is installed worldwide, and 100 GW of PV modules are annually produced to be installed. This situation has resulted in a (2019) price for mainstream PV modules of just 25 ct/watt-peak (PV Magazine 2019). At present, in some sunbelt countries, PV systems can therefore generate electricity at just 3 ct/kWh. This is lower than the cost of electricity produced by a coal plant. It is expected that this trend will continue.

Currently, the efficiency of mainstream PV modules, which for 95% of the market share are made of silicon solar cells, is in the range of 16.5%–18%. This means that such PV modules, under an irradiance of 1000 W/m^2, generate a power of 165–180 W/m^2. Because the detailed functioning of PV cells and PV modules goes beyond the scope of this book, we refer to the Chapter 3 for a short overview of PV materials, PV cells, and PV modules as well as several standard indicators that are common for addressing the performance of PV applications. The indicators are efficiency and nominal power, watt-peak, and performance ratio. More information about the fundamentals of PV technologies can be found in the recently published book *Photovoltaic Solar Energy—From Fundamentals to Applications* (Reinders et al. 2017).

Price reductions and a strong focus on enhancing the efficiency of PV modules was required for successful, broad adaptation of PV systems. Some people say that we even have been too pessimistic about the adaptation of solar technology. Namely, if this trend will continue in the coming years, electricity generated by solar PV systems in 2030 will be at least five times as high as in 2018 (IEA 2019).

To reach the goals of the Paris Agreement signed in 2015 (UN 2015), 100% renewable supply will be necessary by 2050. Therefore, a smooth energy transition from fossil fuels to renewable energy sources is one of the principal challenges that mankind faces at the moment. It will delay and hopefully stop climate change caused by the emission of large quantities of CO_2 and other greenhouse gases. Namely, these emissions largely originate from our fossil energy demand, which as such should be drastically reduced. In 2015, it was therefore agreed upon by the Paris Agreement that the global temperature rise should stay below 2°C, actually preferably below 1.5°C, compared to preindustrial levels. This change from a fossil fuel society to one based on renewable energy will require a huge effort by a varied group of stakeholders: the public, policy makers, scientists, industry, engineers, and designers. Many international and national agreements state that solar PV energy technologies will be one of the main contributors to achieve a prospective 100% renewable energy supply. And it is generally believed that both societal acceptance and technology development will be key to realize this goal. It is known that the interdisciplinary field of design can bring these two aspects together by the creation of products or systems, which people like to use.

As such, it is not only low prices and high efficiencies that will lead to a successful adaption in solar technology. Because solar PV modules can be visually observed in the public space, their image and visual appearance are becoming more important and might need changes (Figure 1.1). They no longer have to be black or blue, or directed at a certain, fixed angle toward the sun. They no longer have to contain a small part of the roof. Solar PV technologies can become a natural part of our environment, our buildings, and our cars. New PV technologies make it possible to create more diversity in placing and integrating solar cells, in terms of orientation and positioning, color, transparency, and even flexibility and form giving. These developments are needed to stimulate the large-scale adaptation of the technology, even more so in cities, where the energy demand density is higher than elsewhere, and space is rare. Since the price of solar electricity has lowered, there exists now an opportunity to consider the aesthetics features of PV modules and PV systems for innovative ways of integration in buildings and landscapes whether they are city or rural landscapes. Luckily, this is already happening in several PV projects.

The fact that citizens seem to be opposed against big solar parks is understandable. Namely, in their perception, beautiful nature landscapes are being filled with PV systems. It is important to be aware of a possible rejection of solar energy because in the recent past, people also weren't supportive to windmills in their surroundings leading to the NIMBY effect, which means Not In My Back Yard. The Netherlands, or Low Countries, is a very flat country where sight goes far. Thanks to well-planned projects, windmills and wind parks have become an icon of the Dutch landscape. Perhaps, if we design them well, something similar will happen with solar PV systems (Figure 1.3). Design will be undeniably necessary to create a new solar revolution beyond the already initiated energy transition.

FIGURE 1.3 A well-integrated PV system looking like a French garden in the Efteling in the Netherlands. (Courtesy of Angèle Reinders.)

1.2 ASPECTS OF INTEREST FOR DESIGNING WITH PHOTOVOLTAICS

Despite the situation sketched in the previous section, at present, the application of PV solar cells and PV systems beyond primary energy production is still limited (Figure 1.4). Earlier work revealed that the design potential of PV solar cells and PV systems is often not fully used (Eggink and Reinders 2016); therefore, in this book, the opportunities and challenges of designing with PV materials and systems are explored in the context of five aspects that are relevant for successful product design.

The five aspects that are relevant for successful product design (Reinders et al. 2012) are: (1) technologies and manufacturing, (2) financial aspects, (3) societal context, (4) human factors, and (5) design and styling, which all together form the so-called innovation flower (Figure 1.5). These five aspects, the context for this book, are subsequently discussed.

Technologies and manufacturing deals with PV materials that are used and the manufacturing techniques that are used to create PV cells and PV modules. Also, the electronic equipment that is applied to convert, distribute, monitor, and store solar energy plays an important role.

The *financial aspects* deal with investments in solar systems and related PV products and the economic value of the energy produced.

(a) (b)

FIGURE 1.4 Two examples of designs that contain PV technologies that have been well integrated: (a) "The Current Window" by Marjan van Aubel made of PV solar cells acting as stained-glass elements, and (b) a solar-powered chandelier by Nina Edwards Anker.

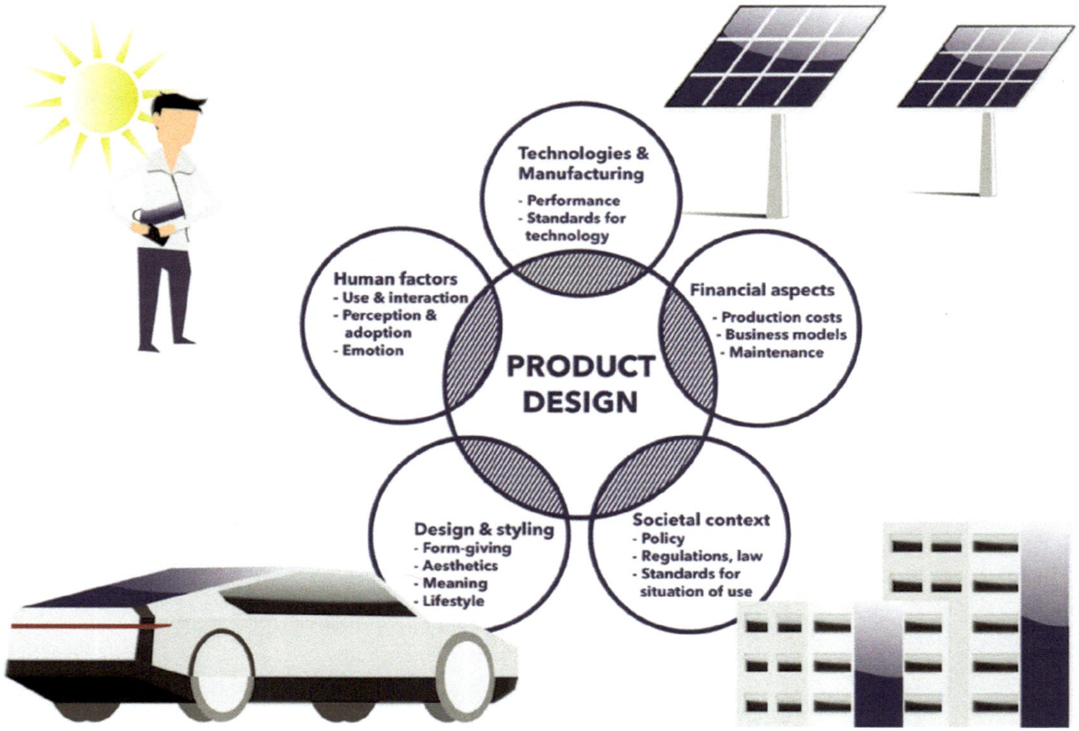

FIGURE 1.5 Innovation flower of industrial product design showing objects that contain integrated solar cells, such as a parka which can power mobile phones, a solar powered electric passenger car, a building with integrated PV facades and conventional PV arrays with tracking systems.

The *societal context* plays an important role in the realization and acceptance of PV systems within society. Policy, regulations, laws, and standards are typically categorized as societal aspects. However, the public opinion on sustainability and the willingness to use PV technologies play important roles here. For instance, from a prior study, it was concluded that respondents were quite positive about PV products that have an environmentally friendly or a social character (Apostolou and Reinders 2016).

The fourth aspect of *human factors* deals with the use aspects of PV systems. This is especially important in the case of PV systems in smart-grid solutions and product-integrated PV. In the case of smart grids, for instance, the fact that energy should be used when it is available, or in moments when the generation costs are low, could compromise the usability of appliances when user interfaces of home energy management systems are not well designed (Obinna et al. 2016).

The last aspect, *design and styling*, deals with the appearance of PV technologies, whether these are stand-alone or integrated in products. An interesting and contemporary appearance can have a major influence on the desirability of PV-integrated products but can also play a role in making PV systems more acceptable by its users or in its environment of use, for instance, in the case of building-integrated PV (BIPV) systems. Well-designed objects tend to encounter less resistance and also have an increased functionality because of a positive and forgiving attitude of the user. As Donald Norman stated, "Attractive things really do work better" (Norman 2004). Moreover, the communication function of design (Crilly et al. 2004) can help to improve on the other four aspects, in particular, the human factors, and hence, stimulate the acceptance of innovative technologies (Mugge and Schoormans 2012; Eggink and Snippert 2017). This can be put in the vision of Sagmeister and Walsch (2019) who run a creative agency in New York. According to them, beauty itself is a function. It has transformative powers, and it influences the way we feel and behave. They believe that beauty can improve people's lives. Beauty means reaching beyond what just works or what is simply pretty, making things more joyful and better functioning.

Usually in the PV industry only two out of these five aspects, namely *technologies and manufacturing* and *financial aspects*, are emphasized in product development of PV modules. Sometimes a third aspect, the *societal context*, is taken along in the design process; however, the two remaining aspects *human factors* and *design and styling*, which are also required to create a successful product in the market, are usually neglected. This is a serious omission from the product development chain, which potentially can negatively affect consumers' long-term interest in PV products and, as such, might limit the full exploitation of all PV markets. Therefore, it makes sense to evaluate how we can design with PVs instead of just technologically applying it (Eggink and Reinders 2016). And in this scope, Industrial Design Engineering can play an important role.

Currently, the role and the impact of Industrial Design Engineering in the Western world is still increasing. Even though not always recognized, Industrial Design can be a key factor in making a company and its products competitive because it improves and strengthens the company's positions in their markets and succeeds in translating technologies into products that have a different, innovative image. Companies that invest in design tend to be more innovative, more profitable, and grow faster than those who do not (EU 2009; DZDesign 1996; Gemser & Leenders 2001). Likewise, this is confirmed by other studies on the Dutch design sector by TNO (2005) and Gemser et al. (2006). These studies clearly state the significant contribution of Industrial Design Engineering to the competitiveness of industry. If this is true for a diverse range of different industries, then this should also apply to the PV sector.

As mentioned before, the aesthetic appeal of PV modules and PV systems will become more important; now they will have greater visibility through their integration in buildings, cityscapes, and rural landscapes. Therefore, attention should be paid to beauty in the context of product design. Also, Steve Jobs subscribed to the idea that aesthetics is as important as functionality (Isaacson 2011). He dedicated his whole career to everyday electronics products that aren't just efficient or useful, but also well-designed. The best-known example resulting from Jobs' visionary ideas is the iPhone, which with a simple, elegant shape has been copied by many other cell phone brands.

1.3 DEFINING INDUSTRIAL DESIGN ENGINEERING

Before continuing, it would be useful to shortly evaluate the definitions that exist for Industrial Design Engineering. For instance, the World Design Organization (WDO 2019), formerly known as the International Council of Societies of Industrial Design (ICSID) (ICSID 2005), defines Industrial Design as "a strategic problem-solving process that drives innovation, builds business success, and leads to a better quality of life through innovative products, systems, services, and experiences." According to an extended description, Industrial Design is "a strategic problem-solving process that drives innovation, builds business success, and leads to a better quality of life through innovative products, systems, services, and experiences. Industrial Design bridges the gap between what is and what's possible. It is a trans-disciplinary profession that harnesses creativity to resolve problems and co-create solutions with the intent of making a product, system, service, experience or a business, better. At its heart, Industrial Design provides a more optimistic way of looking at the future by reframing problems as opportunities. It links innovation, technology, research, business, and customers to provide new value and competitive advantage across economic, social, and environmental spheres" (WDO 2019).

These definitions are very descriptive and rather instrumental, showing more or less what designers do: working in offices while being creative for the benefit of companies, clients, consumers, and users.

A more meaningful definition of what Industrial Design really can bring is needed to understand its role in the engineering process. Looking at the first definition, the "humanization of technologies" can also be referred to as "making technology available for people." Other important aspects in this definition are the creativity and the cultural and economic exchange. On a broader level, design can be seen as the human capacity for changing the world around us in a preferable direction: "design, stripped to its essence, can be defined as the human capacity to shape and make our environment in ways without precedent in nature, to serve our needs and give meaning to our lives" (Heskett 2002). In this context, industrial designs can serve two types of functionality: the *utility* function serving our needs, and the *significance* function providing meaning. The latter is related to the previously mentioned cultural exchange. The economic factor and the importance of innovation in making products and companies competitive mentioned in the previous paragraph is apparent in the definition of design as a creator of value by Dorst (2011). In this definition, design practice is represented with the simple equation "WHAT (thing) + HOW (working principle) leads to VALUE (aspired)." In the equation, the "value" can represent all sorts of things from a more convenient way of delivering packages to a more sustainable way of lighting the streets. The "value" can relate to both economic value for consumers and companies, as well as cultural value for users and society, which is also apparent in the definition by the WDO.

When related to innovation, this dual characteristic of the value of Industrial Design is best illustrated by the theory of *Design-Driven Innovation* by Verganti (2009). Here, innovation can be derived from both technology development—inventing a more efficient way to harvest the energy from the sun—as well as the design of things (Figure 1.6). Verganti speaks of design-driven innovation as "radically changing what things mean," which refers to the previously mentioned cultural value of things. One can think of the design of solar panels, which at first were only seen as technical installations, to be set up on rooftops and in desserts. However recently, solar panels have become available in different colors and prints, changing the meaning of the panels from "installation" into "decoration." In this sense, Verganti distinguishes his theory from existing opposing innovation paradigms of either technology push or market pull or a combination thereof, by the introduction of the concepts of meaning and cultural value.

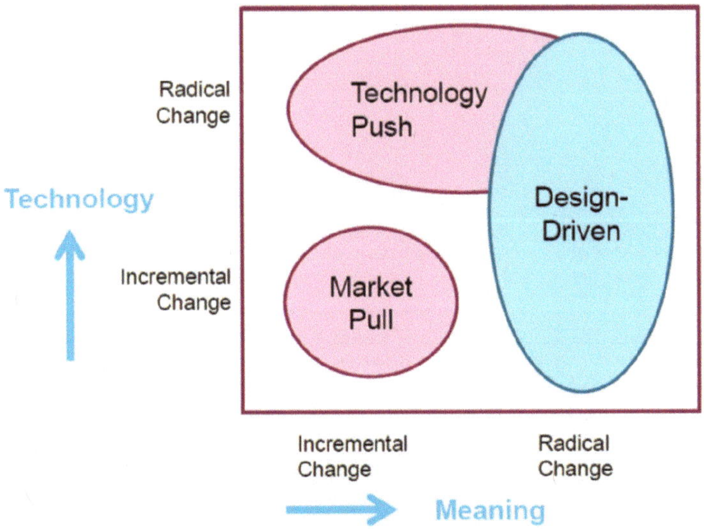

FIGURE 1.6 Design-driven innovations. (Adapted from Verganti, R., 2009, by Eggink and Reinders.)

The dual characteristic of innovation is in its turn also apparent in the characterization of design by Heskett; technological innovation refers to the improvement of the utility function, whereas design-driven innovation refers to the improvement of the significance function. So our definition of design for this book can be written as:

> Design is the creation of value, which can be derived from both the innovative application of working principles (utility) as well as the innovative design of meanings (significance).

Technology also needs to be accepted, and even naturalized. Van Mensvoort (2019) created the Pyramid of Technology. It describes the various levels of functions in peoples' lives (Figure 1.7). "Technology can become so accepted that we experience it as a vital or even a natural part of our lives" said Van Mensvoort in 2014. His pyramid is inspired by Maslow's Hierarchy of Needs (Maslow 1943), which describes human requirements, such as nutrition, shelter, security, and love in subsequent stages. Similar to Maslow's model, technologies can move up and down through various levels of the pyramid, while lower stages need to be fulfilled before the next stage can be attained. It can serve as a tool for scientists, inventors, engineers, designers, and entrepreneurs to position themselves in the playing field of technological development and eventually create better technologies, products, or systems.

A new technology may seem artificial at first (e.g., quantum computing). But when the technology rises from the base of the pyramid toward the top, it can become so accepted that we experience it as a vital or even natural part of our lives. We could say that at the moment PV technologies are applied and accepted; however, they definitively have not been naturalized or become invisible yet.

According to Dorst (2011) again, the complexity within contemporary conceptual design is that in most cases only the desired value of the end result is known at the

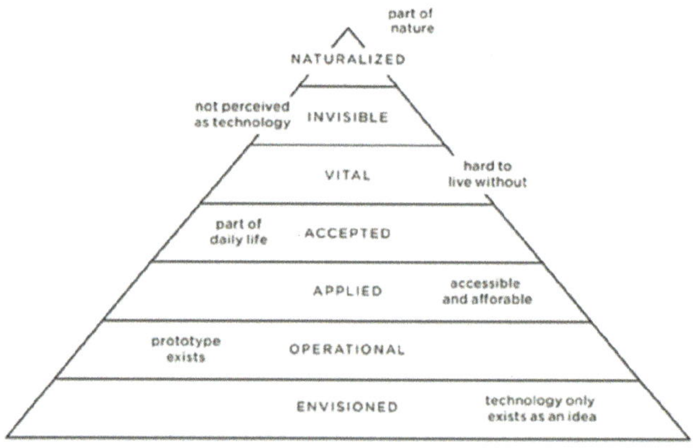

FIGURE 1.7 Pyramid of Technology by Koert Van Mensvoort.

beginning of the process. This means that design engineers, in order to reach considerable value, must develop the "what" (a design, product, or service) and the "how" (a working principle) simultaneously. To cater this creative problem-solving process, one can see that it is important to know as many working principles as possible. "What," in its turn, leads to the purpose of this book: informing our readers about the many possibilities of PV technologies as a "how" for design.

1.4 STRUCTURE OF THIS BOOK

Given the framework of the innovation flower and context of Industrial Design Engineering, the setup of this book is as following. In Chapter 2, a short history of PV products will be given. In Chapter 3, PV technologies and their design features related to—among others—coloring and form giving, are presented to provide a general context for the subsequent chapters about product-integrated PV (Chapter 4), BIPV (Chapter 5), and users of PV products (Chapter 6). In between these chapters, various interesting design cases are shown, which will give insights in actual design processes as well as in challenges encountered and solutions found by the design of a solar-powered table; the colorful products that can be designed by luminescent solar-concentrator PV technologies; the design of a solar-powered chandelier called Virtue of Blue; solar racing cars; the Solaris building, which is a BIPV project; and a solar-powered charging station for e-bikes.

REFERENCES

Apostolou, G. and A.H.M.E. Reinders (2016). How do users interact with PV-powered products? Investigating 100 lead users and 6 PV products. *Journal of Design Research*, 14(1), 66–93.

Crilly, N., J. Moultrie, and P.J. Clarkson (2004). Seeing things: Consumer response to the visual domain in product design. *Design Studies*, 25(6), 547–577.

Dorst, K. (2011). The core of "design thinking" and its application. *Design Studies*, 32(6), 521–532.

DZDesign (1996). *Diseno industrial: Beneficio para las empresas*. Bilbao: DZDesign.

Eggink, W. and A.H.M.E. Reinders (2016). Product integrated PV: The future is design and styling, *Proceedings of 32nd European Photovoltaic Solar Energy Conference and Exhibition*. Munich, Germany: EU PVSEC.

Eggink, W. and J. Snippert (2017). Future aesthetics of technology; context specific theories from design and philosophy of technology. *The Design Journal, Design for Next: Proceedings of the 12th European Academy of Design Conference*, Rome, Italy, pp. 196–208.

EU (2009). Design as driver of user-centred innovation. C. o. t. E. Communities. Brussels: Commission of the European Communities.

FhG-ISE (2019). Fraunhofer Institute of Solar Energy Systems, Photovoltaics Report, www.ise.fraunhofer.de, visited June 2019.

Gemser, G. and M.A.A.M. Leenders (2001). How integrating industrial design in the product development process impacts on company performance. *Journal of Product Innovation Management*, 18(1), 1–61.

Gemser, G., D. Jacobs, and R. Ten Cate (2006). Design and competitive advantage in technology-driven sectors: The role of usability and aesthetics in Dutch IT companies, *Technology Analysis & Strategic Management*, 18(5), 561–580.

Heskett, J. (2002). *Toothpicks & Logos, Design in Everyday Life*. Oxford: Oxford University Press.

ICSID (2005). Definition of Industrial Design. http://www.icsid.org/about/Definition_of_Design/. Accessed May 2005.

IEA (2019). International Energy Agency, Tracking Clean Energy Progress: Solar PV, www.iea.org/tcep/power/renewables/solarpv/, visited June 2019.

Isaacson, W. (2011). *Steve Jobs: The Exclusive Biography, Little*. London, UK: Brown Book Group.

Maslow, A.H. (1943). A theory of human motivation, *Psychological Review*, 50, 370.

Mugge, R. and J.P.L. Schoormans (2012). Product design and apparent usability. The influence of novelty in product appearance. *Applied Ergonomics*, 43, 1081–1088.

Norman, D.A. (2004). *Emotional Design: Why We Love (or Hate) Everyday Things*. New York: Basic Books.

Obinna, U., P. Joore, L. Wauben, and A.H.M.E. Reinders (2016). Insights from stakeholders of five residential smart grid pilot projects in the Netherlands. *Smart Grid and Renewable Energy*, 7, 1–15.

PV Magazine (2019). Module Price Index, May 2019, www.pv-magazine.com/module-price-index/, visited June 2019.

Reinders, A.H.M.E., J.C. Diehl, and J.C. Brezet (2012). *The Power of Design*. Chichester, UK: Wiley, p. 331.

Reinders, A.H.M.E., P.J. Verlinden, W.G.J.H.M. Van Sark, and A. Freundlich (2017). *Photovoltaic Solar Energy from Fundamentals to Applications*. London, UK: John Wiley & Sons.

Sagmeister, S. and J. Walsh (2019). *Sagmeister & Walsh: Beauty*. New York: Phaidon Press. ISBN: 9780714877273. https://sagmeisterwalsh.com/about/, visited in June 2019.

TNO (2005). *Design in the Creative Economy – A Summary*. Amsterdam, the Netherlands: Premsela Foundation.

United Nations (UN) (2015). Paris Agreement. https://unfccc.int/sites/default/files/english_paris_agreement.pdf.

Van Mensvoort, K. (2019). *Next Nature-Waarom technologie onze natuurlijke toekomst is*. Amsterdam, the Netherlands: Maven Publishing.

Verganti, R. (2009). *Design-Driven Innovation, Changing the Rules of Competition by Radically Innovating What Things Mean*. Boston, MA: Harvard Business Press.

WDO (2019). World Design Organization (WDO), Definition of Industrial Design. https://wdo.org/about/definition/. Accessed June 2019.

Design Case 1

Current Table

Marjan van Aubel, Peter Krige, and Angèle Reinders

1.1 INTRODUCTION

Current Table is a connected, self-powered table with integrated dye-sensitized solar cells (DSSCs). The cells are a third-generation solar technology and are so sensitive to light that they work even in diffuse light, both natural and artificial. Their sensitivity and responsiveness to light opens up the potential for solar products and applications beyond the traditional PV direct sunlight model.

This solar innovation inspired the creation of Current Table and with a strong design approach led to some unique qualities of the integrated DSSC modules.

The table is seen as a "living object." It is autonomous and communicates intelligently to the user where it functions best. There are embedded light sensors that measure the light intensity and spectrum components in the room, which are visualized through a mobile application.

Away from any mains wired infrastructure, devices such as tablets and phones can be charged through integrated batteries and USB ports in the table legs. The table is an example of an object with a double function, both a functional power source and a subtle, aesthetic object where solar technology is integrated naturally.

1.2 VISION: LIVING OBJECTS

How can solar technology be more integrated into our daily environment? The decrease in price and increase in efficiency of solar technology has not been enough to ensure adoption by consumers, homeowners, and cityscapes. Rules and regulations can prevent access to these technologies. For someone who doesn't own their own house, or lives in an old city such as Amsterdam, it is often not permitted to have solar technology on their roof because it will influence the protected architecture. How can these regulations be adapted or influenced? Schemes such as renting solar panels on a different location would make it possible to own or use solar panels. Again, the relationship the consumer has with solar technology is even farther away. If the consumer is confronted with this on a daily basis, it will also affect his or her awareness and behavior.

The relation between energy production and the objects that use it needs to be reconsidered. True behavioral change is needed for the adoption of a cleaner technology. To do this, design plays a significant role; a sensitivity to the aesthetics around us is required.

Design enables objects to inhabit our spaces and become embedded into our daily lives and experiences. What if everyday objects turn into power sources

FIGURE DC-1.1 Current Window, 2015 London.

and power themselves? A window that doesn't only work as window but can also work as a power source that generates electricity.

"Current Window" designed by Dutch designer Marjan van Aubel (Figure DC-1.1) is a modern version of stained glass that generates its own electricity. Colored dye cells in the window itself use the properties of color to generate useful energy. It also forms part of a "smart" system, working with other connected home systems to provide additional energy. Through integrated USB ports, devices can be charged directly from the window sill or be connected to charge a centralized home battery.

Alongside the window is another living object: "Current Table" (Figure DC-1.2). It uses built-in solar cells to generate its own electricity. Both objects use transparent "organic" and "dye-sensitized" solar cells to make electricity. These are both solar technologies that mimic photosynthesis to create an electric current. The aim is to give power to the consumer. Use the possibilities of our environment, the sun, and transform that into usable energy through design.

FIGURE DC-1.2 Current Table, 2017 London.

The table and the window are the first applications of a bigger vision: *connected living objects*. These are self-powering objects that create a local low power output while interacting with the environment and its inhabitants. The technology developed for this table can be used in different applications: facades, windows, and many other small objects. Producing local off-the-grid power is a possible future. By starting small, it is a way to redefine solar technology and make it a natural part of the environment.

The integration of solar cells into a product, or any energy-generating technology, is a key component in creating a living object. By adding intelligence through sensors, computer processing, and connectivity they become connected living objects and become transmitters of a larger idea for the consumers to be active and aware energy providers or participate in generating energy for their needs.

The overall vision for the Current Table was to create a product that should harvest solar energy using embedded solar cells and make it available for people to use. It should be something that is integrated naturally into people's lives but also increases the aesthetic appeal and consumer desirability of solar technology. The use of dye-sensitized solar cells (DSSCs), which have the potential for customizing the properties of color, pattern, and transparency, should provide beauty through novel patterns. Not only should the cells add to the beauty, but also the table's structure itself should be inspired by biomimicry, referencing plants and photosynthesis, to create a unified design. This design and aesthetic could then capture and embody the idea of a living object and possibly inspire our relationship to energy, light, and ultimately our connection to the sun.

1.3 DSSC: ORGANIC SOLAR CELL TECHNOLOGY

A DSSC cell is made from two pieces of glass coated with a transparent conductive film (Figure DC-1.3). These face toward each other and act as electrodes. On the surface of one of the electrodes, a layer of titanium dioxide particles is attached, which is coated with a sensitized dye required to create excited electrons from incoming photons, and on the other electrode, very thin film of platinum is attached, which acts as a catalyst in this system. The electrolyte in the cell serves, among others, as a charge transport medium.

When a DSSC is exposed to irradiance, photons can be absorbed by the dye. This causes excitations of electrons, which can become free charges that can move through the titanium-dioxide layer into the so-called photoelectrode. Next, these electrons flow through the external circuit while completing some form of work on their way to the counterelectrode, such as powering a light bulb or charging a battery. Subsequently, a chemical (redox) reaction happens at the surface of the electrode, and the electrons are then transported by the electrolyte back to the dye. The repetition of this process is what allows DSSCs to capture energy from the sun at an overall peak power conversion efficiency of 11%–15%.

One of the main advantages of DSSCs is that they still function at a reasonable efficiency under diffuse and weak indoor light. This makes DSSCs a viable option for powering low-energy devices in the house. In addition, their transparent appearance allows them to be integrated where traditional solar panels could not, such as in a window.

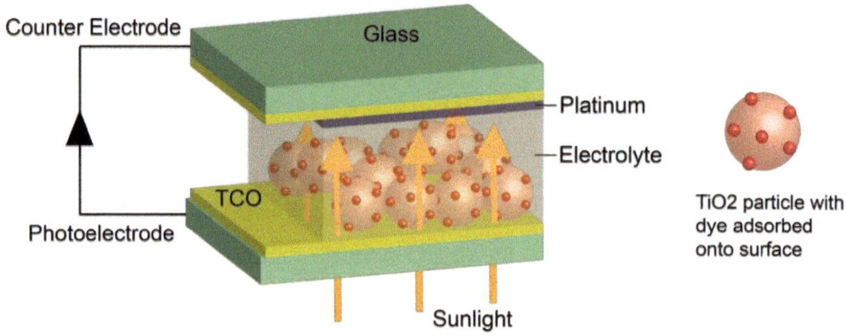

FIGURE DC-1.3 Cross section of a DSSC.

A downside of DSSCs is that the cells' materials can degrade when they are exposed to UV light. In addition, the use of the liquid electrolyte could be a disadvantage in cold surroundings, namely, at low temperatures the electrolyte can freeze, leading to breakage and leakage of the cells.

1.4 COLOR AND EFFICIENCY: FUNCTIONAL AND VISUAL CONSIDERATIONS

One of the greatest things about DSSCs is that they are colored. This holds significant advantages over the black and blue gridded rectangles in conventional silicon solar panels because it allows the designer to start thinking about the integration of solar as a design feature rather than an afterthought. The option of more variability means that the styling of the solar cells can be fitted to and made more appropriate for the context of the specific application. However, the coloring of DSSCs does not come without its limitations. A variety of colors can be used, although different colors have different efficiencies depending on their place in the color spectrum, with dark red being the most efficient because it is closest to infrared (Table DC-1.1).

This means that compromises must be made when designing with these solar cells in order to maintain a high enough output. Even with this constraint, interesting designs can still be made, and it's even possible to mix colors on a single

TABLE DC-1.1
Varying performance of 30 × 30 cm DSSCs according to their color

DSSC module	I_{sc} [mA]	V_{oc} [V]	FF [%]	P_{max} [W]
Red	570.1	9.9	47.3	2.6
Orange	407.9	10.0	53.2	2.1
Light orange	323.2	10.1	54.8	1.8
Green	234.8	7.9	56.7	1.0

tile. Orange was chosen for the Current Table because of its best performance and stability indoors (Figure DC-1.4).

Green was the most favored during the design process; however, its low efficiency excluded it from the intended application, which needed a minimum power density (Figure DC-1.5).

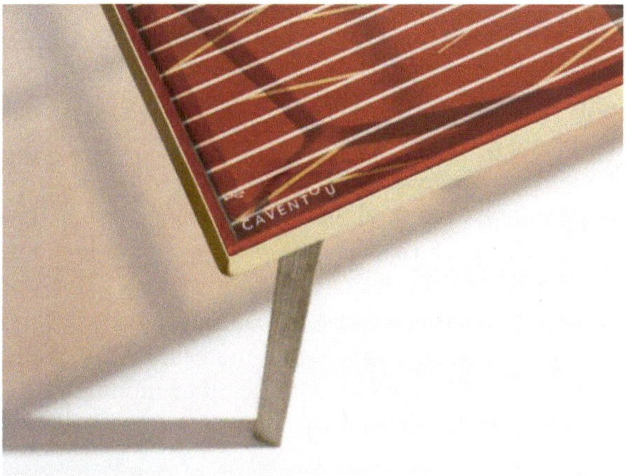

FIGURE DC-1.4 Caventou table in orange.

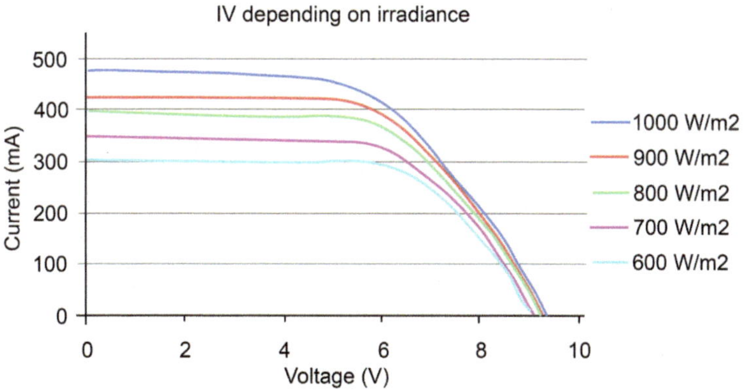

Irradiance (W/m²)	I_{sc} [mA]	V_{OC} [V]	P_{MAX} [W]	I_{MAX} [mA]	V_{MAX} [V]
1000	470.7	9.34	2.47	404.8	6.10
900	424.5	9.22	2.33	374.7	6.21
800	393.2	9.18	2.20	352.8	6.24
700	343.6	9.10	1.98	314.5	6.28
600	296.6	9.00	1.74	268.3	6.48

FIGURE DC-1.5 Graph and table showing how DSSC performance varies with irradiance.

1.5 AESTHETICS OF DSSC: MATERIALS AND PATTERNS

To design a table with integrated DSSC solar cells, one must take into account multiple conflicting criteria. Normally, manufacturing processes and the functional construction components are highlighted with an emphasis on efficiency and cost. However, one also needs to consider aesthetics, attractiveness, and consumer desirability. Thus, it is important that the aesthetic value increases. This can be at the expense of efficiency, as long as it be relatively small or appropriate to the application. In terms of a more holistic approach to solar design, aesthetics and beauty then become an important criterion competing with the efficiency, cost, and stability of the cells.

One can apply this holistic approach to a single DSSC module, creating a good design on its own and in combination with adjacent modules or other interfacing or overlapping materials in the design. Because solar cells are often connected in series or parallel to other cells, the design can be seen as part of a larger tiling or repeating pattern of modules. Therefore, each cell needs to make sense in relation to other modules. Another option for bringing the modules together as one is to arrange the modules underneath a large cover glass or optically transparent material. This can have a pattern printed on top that can add to or connect the other modules together visually (Figure DC-1.6).

FIGURE DC-1.6 Pattern design Current Table.

1.5.1 Functional Constraints

Having something made by a manufacturing partner on the other side of the world brings with it further constraints. These act as a reality check for the designer and bring the blue-sky thinking of solar design to the ground. Due to manufacturers having strict processes already in place, the designer must design in accordance with these. However, there is always space to push boundaries. Proper feedback and reflection by the design team on the manufacturing processes and component design, which requires a close collaboration with the manufacturing facilities and manufacturing teams involved, a better component can be produced and, ultimately, a more tailored component for designers to work with and integrate into the world.

In the case of working with DSSCs, there are many decisions that have been made by the manufacturer for reasons to do with functionality and efficiency that affect how the cell looks. This includes the size, shape, the number of dye stripes on the cell and the width between them. Stripes must have consistent widths and must be in shades of orange, red, and green for high efficiency.

This specification makes a cell that is very efficient but lacks in aesthetic appeal. On the other hand, this cell could be seen as a blank canvas for a designer to create an artwork, and thus making it desirable. On the blank canvas of the cell, there are an even number of strips. Each of these rectangles can be altered with breaks in them or patches of different colored dye, but too much alteration will lead to an ineffective solar cell. It is a balancing act between good looks and a reduction in efficiency because there is no point in creating an artistic piece if you can't get any power from it.

1.5.2 Creative Response to the Constraints

The design for the pattern on Current Table was heavily inspired by nature. This came from the links between solar cells and photosynthesis along with the bigger idea of "living objects." The final design took from the veins in leaves and had minimalist adornments. This minimal look was dictated by the need for good efficiency (Figure DC-1.7).

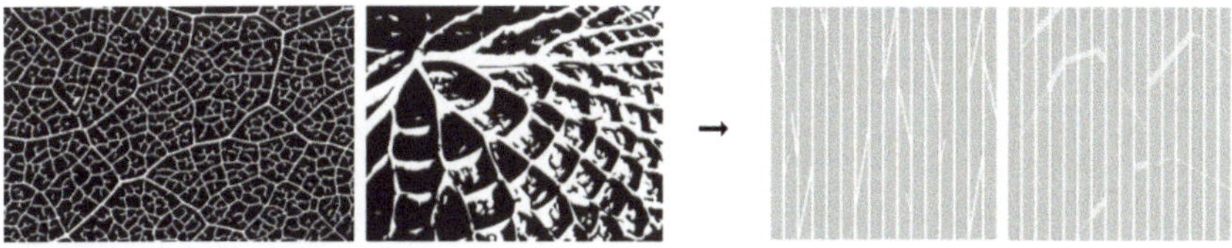

FIGURE DC-1.7 Inspiration pattern design Current Table.

1.6 CELL DESIGN: MODULARITY AND TILING

1.6.1 Introduction to Manufacturers and Modularity

When designing with solar cells it is important to keep the bigger picture in mind. Due to constraints on the size of cells that can be produced, with small cells not giving enough power and large cells being physically impractical, manufacturers tend to modularize what they make. This means that there could be a set size that a cell is made and the intention is for multiple tiles to be used to build a larger surface area. It is the job of the designer to work creatively within these set standards in order to see how the cells could fit into the larger picture for the application.

If a large overall design is being created, one may be held back by cost. This can be due to constraints from manufacturing, meaning that making a pattern from many different designed tiles is not always economically feasible. A creative response to this could be to use a single tile that has a pattern on it that makes sense in a variety of orientations. Here, the designer creates their own aesthetic system that complements the modularity constraints set by manufacturers (Figure DC-1.8).

A simple repeating tile design that can be configured into multiple patterns can be of benefit when designing with solar cells because it makes it scalable and gives a variety of options. This was the case with the Current Table, and as mentioned previously, the importance of the cell pattern being repeatable if at a high variety of ways was very high. This allows the Current Table to be configured in different sizes.

Not only does the Current Table have a modularity in how the individual cells join but also in how the hardware works through the grouping of cells together with the smart corners for charging. For every smart corner, there can be six to nine cells, giving the option of a six-cell table with one smart corner or an 18-cell table with two smart corners (Figure DC-1.9).

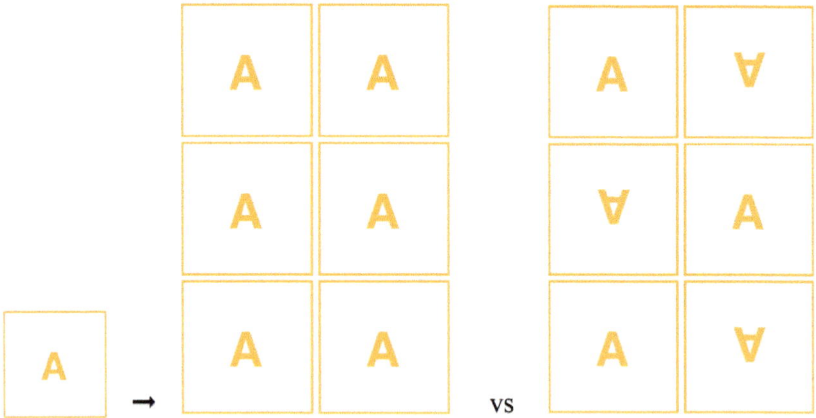

FIGURE DC-1.8 Pattern layout Current Table.

FIGURE DC-1.9 Full-pattern design Current Table.

Having a modular system also aids the fixing of the table if problems with the cells arise. One would simply remove a single tile and replace it with a new one without affecting the system as a whole. On the Current Table, this is aided by the electrical contacts on each cell being at the four corners.

1.7 TABLE DESIGN: MATERIALS AND STRUCTURE

The design of the table is made in such a way that it communicates itself visually as a living object. The thin aluminum frame lends itself to be as slim as possible and holds the glass surface on the very edge. This allows as much light as possible, which creates a maximum surface area for the solar panel.

The legs, based on biomimicry,[1] are shaped into a structure that reflects how a tree branches out. The orange tabletop references leaves and cells at the

[1] The design and production of materials, structures, and systems that are modeled on biological entities and processes.

FIGURE DC-1.10 Tree branches.

microscopic level, and the vein-like pattern shows electrons bouncing from one cell to another. Corners naturally branch out to give users energy. Here the USB port allows users to charge their phones or tablets (Figures DC-1.10 and DC-1.11).

The design of the table legs also highlight another aspect of modularity in that they stay the same whilst the design could get smaller or bigger with the number of cells. This size must make sense at both a human scale and solar scale. It must fit within buildings but be large enough to collect enough sunlight to power the smart-corner charging points.

1.7.1 Reflection on Selected Materials

If this product was to be made for market, there would be much to reflect on with the design. Whilst embedding new, renewable technologies into everyday objects is no doubt the way forward, these objects must align with what consumers want. For instance, glass and metal tables do feel cold and are hard to keep clean.

1.7.1.1 Light Mapping: Perceiving the Environment

The Current Table also communicates itself literally as a living object in that it is able to perceive the environment intelligently through an array of light sensors integrated into its surface. This allows the table to experience an interior environment in the same way a living being might experience it, like biomimicry (Figure DC-1.12).

FIGURE DC-1.11 Branch structure table legs.

The Current Table app also communicates to the user about how the energy stored in the integrated batteries in an understandable, human way. Instead of only using energy units to state stored chemical energy and current solar power available, it relates it to smart-device charges. Thus, a user can quickly see if there is enough solar energy or previously stored capacity to charge their particular device.

The table takes this experience and combines it with an innate knowledge of light and interior design, light and human health, and light and behavior. On a more practical level, the light sensors provide a direct feedback to the user about the table's position and its ability to harvest solar energy. This design feature is elegantly communicated visually through graphics and color on a mobile application, making for a more intuitive experience of how light influences the performance of the DSSC modules (Figure DC-1.13).

This built in intelligence is further extended through cloud infrastructure. Once the data is retrieved by the app via a Bluetooth connection to the mobile device, the data is pushed to cloud services, where it could be further analyzed on the internet.

1.7.1.2 Low-Power Electronics: Ensuring Efficiency

There is a concept in LED technology called efficacy, which describes the relationship of light emitted to power used generating it. It is the ratio of luminous flux to power, measured in lumens per watt. This measure is critical to the design of efficient light sources.

FIGURE DC-1.12 Caventou app.

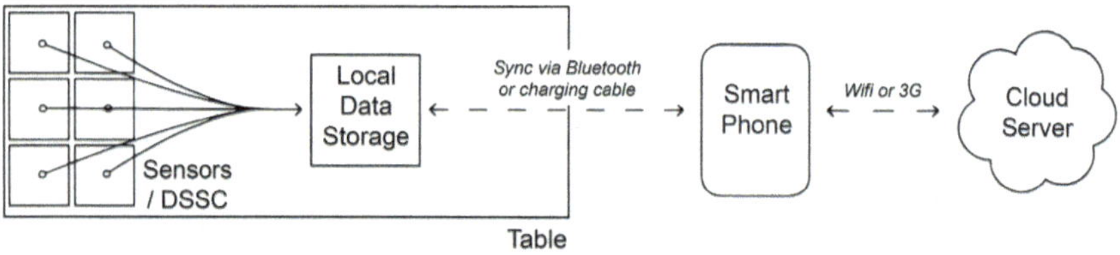

FIGURE DC-1.13 Radio Bluetooth connection.

This idea was taken into consideration when developing the overall electronic system controlling the Current Table, beyond solar-energy harvesting. Because the table had additional features and intelligence, it had to allocate energy resources outside of storing and charging devices via the USB ports.

The light sensors, Bluetooth radio, current sensors, solar-charge controller, and LED display all needed to be highly efficient and low powered.

Again, as with visual aesthetic features enhancing the design of the DSSC module, these design features improved the overall experience of the device.

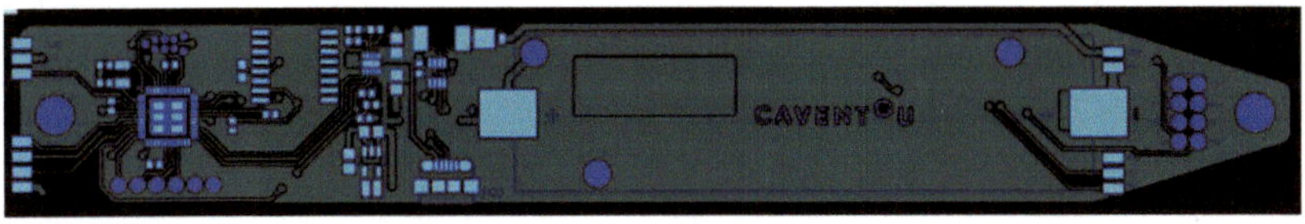

FIGURE DC-1.14 PCB Current Table.

The potential for this information to enhance the product experience and user engagement outweighed the small drop in overall efficiency as a solar charger (that needs as much as possible of the solar energy harvested and stored to be available for charging external devices) (Figure DC1.14).

1.8 CONCLUSIONS

Although self-powered products have been seen in the past as a drop in the ocean of energy usage, there is still a lot of potential in autonomous objects. New lifestyles make our lives mobile, where we can work anywhere and from any place, which also needs a more mobile energy approach.

A combination of large solar and wind parks at sea, away from our scarce landscape but close to existing energy infrastructures, together with direct-energy harvesting in the urban environment is the way to go. Facades, roads, and roofs can be used as potential energy surfaces together with smaller applications where both energy harvesting and function are integrated into one system. The Current Table as a Living Object performs both functions—both a work surface and the energy source to work from. Because these objects live in our daily environment, design plays a significant role because we are surrounded by these objects on a daily basis. Prices of solar PV technologies have dropped, and efficiency has increased, so there is now room (and need!) for aesthetics in PV. A collaboration between both the design/architectural field and the PV industry is needed to have solar integrated in our built environment.

To speed up this awareness and adaptation, solar also needs to become more accessible and desirable in order for people to change their behavior. The Current Table investigated this through the production of a low-power energy device. Additional intelligent features such as lightmapping and charge cycle status were added to enhance the user experience and raise user awareness of solar energy.

This information was presented through a mobile app and LED display in the table legs using a Bluetooth radio, light, and current sensors. All these features needed to be highly efficient and low powered. The table uses transparent colored DSSCs because this technology is suited best for diffuse light. To design the Current Table, different factors have been taken into account, such as functionality, efficiency, aesthetics, modularity, and manufacturability.

At its current state, it can only power tablets and phones. However, the development of both energy capacity and storage of devices develops rapidly.

FIGURE DC-1.15 Current Table in use.

Taking this into account, plus expanding this concept to larger surfaces and more objects, there is a chance this could lead to an autonomous future (Figure DC-1.15).

ACKNOWLEDGMENTS

Thanks to Jon Williams, Samira Ann Qassim, Instrument Industries, Aran Dasan, Imperial College London, and Alicja Pytlewska.

2 A Short History of Photovoltaic-Powered Products

Angèle Reinders and Wouter Eggink

CONTENTS

2.1 THE VERY FIRST PHOTOVOLTAIC-POWERED PRODUCT

It took over a century since the discovery of the photovoltaic (PV) effect—the conversion of photons to electricity by Becquerel (1839) before solar cells were developed. In the 1950s, PV solar cells were developed at Bell Telephone Laboratories in the USA with the purpose to apply them in products that lacked permanent electricity supply from the mains (Chapin et al. 1954). The solar cells were called silicon solar-energy converters, commonly known as the Bell Solar Battery (Prince 1955). Furnas (1957) reports "The Bell Telephone Laboratories have recently applied their findings in the transistor art to making a photovoltaic cell for power purposes. {….} and exposed to the sun, a potential of a few volts is obtained and the electrical energy so produced can be used directly or stored up in a conventional storage battery. {….} The Bell System is now experimenting with these devices for supplying current for telephone repeaters in a test circuit in Georgia. As to cost, one radio company has produced a power pack using this type of photoelectric cell for one of its small transistorized radios." The *Journal of the Franklin Institute* (Anonymous 1955) mentioned already in 1955 that "these (solar) batteries can be used as power supplies for low-power portable radio and similar equipment" (Figure 2.1).

Expectations regarding the applicability of PV cells in products were high; Sillcox (1955) reports on predictions by researchers of New York University "that small household appliances like toasters, heaters or mixers using the sun's energy might be in fairly widespread use within the next five years (i.e., 1960)." These predictions have not become reality because of the incompatibility of power production by solar

(a) (b)

FIGURE 2.1 Solar-powered radio originating from the sixties: (a) front view, slightly damaged, and (b) top view with integrated solar cells showing the circular wafers that were used at that time. (Courtesy of Larry Kazmerski, photos by Angèle Reinders, 2017.)

cells and the high energy demand of thermal appliances such as toasters, and at that time, the cost of silicon PV cells were about $200 per Watt for high-efficiency cells (Prince 1959). At that time, 12% was considered a high efficiency—and the costs of a dry cell to operate a radio for about 100 hours would be less than a dollar (Furnas 1957). Therefore, it was believed that the solar battery could be an economical source for all except the most special purposes. As such, by the end of the 1950s, silicon PV solar cells were applied as a power supply for satellites (Zahl and Ziegler 1960). The Vanguard 1 satellite (see Figure 2.2), launched in 1958, was the first satellite ever equipped with a solar PV power system, and it announced a new area of space technology with solar-powered satellites. Still at present, the Vanguard 1 is the oldest satellite in orbit.

(a) (b)

FIGURE 2.2 Prototype of the Vanguard 1 satellite that has been used for testing: (a) the satellite, and (b) the interior electronics of the satellite. (Courtesy of Larry Kazmerski, photos by Angèle Reinders, 2017.)

At present, various applications of PV solar cells exist ranging from the previously mentioned satellites, stand-alone PV systems, grid-connected PV, building-integrated PV (BIPV) systems and very large systems with a power of tens of megawatts. The latest category is product-integrated PV (PIPV), which can be applied in consumer products, lighting products, boats, vehicles, business-to-business applications, and arts (Reinders and Van Sark 2012).

2.2 THE DEVELOPMENT OF DESIGNED PV PRODUCTS IN THE COURSE OF TIME

From the very first solar-powered products to the present day, the designs of PIPVs have gradually developed. However, this development started very slowly; for about two decades, calculators were the most dominant PV-powered product on the market (Figure 2.3).

When we look at the timeline, we can notice that the technical appearance of PV-powered products hardly changes for a long time. Only after 2010 we saw radically new designs emerge when solar chargers that resemble trees or flowers came to the market.

Regardless, the solar calculators were received with great enthusiasm in the 1970s. This enthusiasm was mainly based on technical features that were enabled by PV solar cells and the effect of making the owner feel modern. For instance, the journal *New Scientist* of July 20, 1978, reported, "The concerned environmentalist can now calculate the downfall of society without eating into the world's resources in the process. A new calculator that is coming on to the market ....{..}...doesn't need an on/off switch because the power comes from a small panel of solar cells." (Anonymous 1978).

The reference to the "concerned environmentalist" is, however, not visible in the straightforward, rather functionalist—or technical—appearance of the products. When we compare these solar designs with other product designs from the same period (Figure 2.4), we can observe that the appearance is rather lagging behind. The solar calculators of the 1990s still look like the Braun ET22 calculator from 1976, and the integrated chargers of the 2000s are all alike. One might try to argue that this is due to the "timeless" nature of the Braun design language, as is illustrated by the recent rerelease of the ET66 calculator from 1987 that is also pictured here (Braun 2019). However, the transparent solar charger in the 2011–2015 period mimics the iMAC from the 1990s and even the most contemporary tree-like chargers are following the nature-inspired form languages of the pictured lamp and chair from ten years before. This indicates that the potential to attract the future user of PV technologies is not yet fully capitalized.

The tree-like solar chargers are obviously designed to resonate with the user the idea of nature and the associated concept of "sustainability." In the broader design discipline, since the 1970s, it is already common to experiment with such meanings and emotions in the design of products (Eggink 2009). For instance, the concept of sustainability is incorporated by the choice of materials—such as in the cork of the pictured TV design by Starck, the solar cells in the Solar Chandelier, and the recycled fridges of the 3D-printed Endless Chair. In other occasions, the sustainability is represented by the organic shaping of elements of a product—such as the Bone Chair by Laarman or the coral-inspired lamp by Grossman. These approaches

FIGURE 2.3 Timeline of PIPV designs from the 1970s onward. (From Eggink and Reinders 2016.)

can be translated to PV products to make them more attractive. Moreover, integrated approaches, such as in the design by Howard, are reflected in the solar-powered lamps that are also chargers at the same time. This will also decrease the persistent gadget image of PV-powered products and makes them aesthetically pleasing as well. In this context, we like to zoom in to the development of solar-powered watches in the course of time.

FIGURE 2.4 Timeline of comparable product designs from the 1970s onward. (From Eggink and Reinders 2016.)

2.3 THE DEVELOPMENT OF SOLAR POWERED WATCHES IN COURSE OF TIME

In 1971, the world's first solar-cell-powered watch was made by the US firm Uranus. It fancied a distinctive sunburst-like watch face with eight specially developed fan-like solar cells that showed a resemblance to the semicircle-shaped cells of the solar-powered radio described in Section 2.1. To emphasize the innovativeness of the device, the watch had a small digital led display (Figure 2.5a), a technology that was introduced just before by the large watch companies Omega and Pulsar.

(a) (b)

FIGURE 2.5 The first-generation solar-powered watches: (a) Uranus prototype from 1971, and (b) the functionality of the Synchronar solar watch is explained in the advertisement (Bielefeld, 2007). It states: "The watch for tomorrow for the people of today."

Unfortunately, the watch stayed just a prototype (Bielefeld 2007). The relatively large solar cells that were necessary to power the first generations of digital watches caused the designers problems in these days. The Ragen–Synchronar watches that were the first to actually come to the market in 1974 solved the problem by reserving the whole watch face for the solar cells and placing the display hidden at the side (Figure 2.5b). Interestingly, the technologically advanced watches were marketed toward a female audience (The Led Watch 2008) (Figure 2.6).

The Nepro "Solar Solid State" produced by Nepro Watch of Neuchâtel, Switzerland, which was shown to the general public at the 1975 Basle Watch Fair, also had a strange

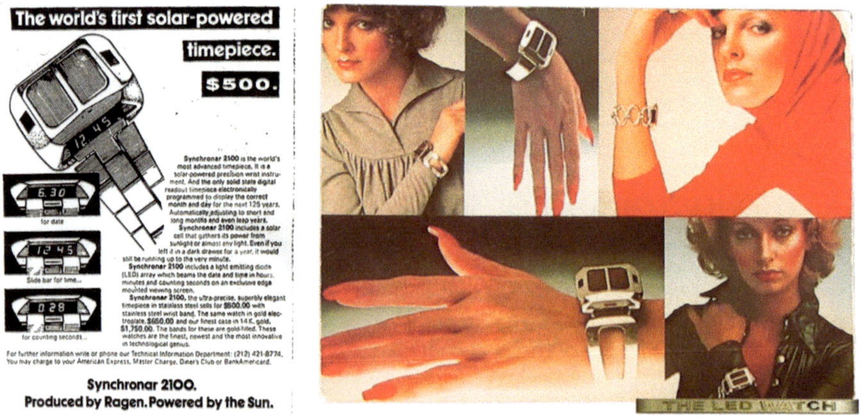

FIGURE 2.6 Two advertisements for the 1974 Synchronar Solar watch by Ragen.

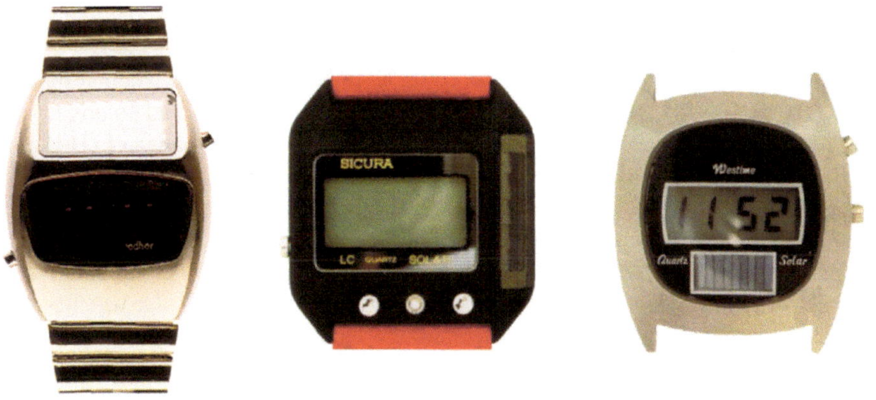

(a) (b) (c)

FIGURE 2.7 The advancement of solar cell technology is visible in watch designs: (a) the Nepro "Solar Solid State" from 1975 (Doensen 1994), (b) the Sicura "LC Solar" from 1978, and (c) the Westime "Quartz Solar" from the same year (Bielefeld 2007).

asymmetric housing to incorporate the huge solar cell (Doensen 1994). Only when the solar cells were more efficient and the watches moved to more energy-saving LCD technology did the designs of solar-powered watches become more mainstream (Figure 2.7).

When Pulsar introduced the "2001" in 1984, the cells were even hardly visible anymore in the background of the face (Figure 2.8a). Because the cells were no longer causing the watch to look different, the innovativeness of the technology was emphasized by the name "2001," a common practice at that time, when everything associated with the upcoming new millennium was regarded as futuristic. With the introduction of the Junghans "Solar 1" costing just DM149, that same year, the price of the solar watch technology was also democratized. In order to still put attention to

(a) (b) (c)

FIGURE 2.8 Analog solar watches with the cells "hidden" in the dial: (a) the Pulsar "2001" (Doensen 1994), (b) and (c) two versions of the Junghans "Solar 1."

the rather hidden solar technology, the later model of the "Solar 1" carried the words "high tech" (Figure 2.8c).

The popularization of solar watches could be rendered complete when at the end of the 1980s, digital watches appeared on the market that mimicked solar cells on the face, while underneath they were still powered by coin cell batteries (Bielefeld 2007) (Figure 2.9).

The increased efficiency of solar cell technology made it possible to implement even more electronics and functionality, which led to the introduction of the "Mega Solar" with radio-controlled time in 1993 (Figure 2.10a). Meanwhile, the integration of solar cells in watches was so widespread that the marketing of solar watches was no longer targeted at the advanced technology but only on convenience: no more battery changing. The friendly priced Casio watches of the time prominently quoted "batteryless" on the dial (Figure 2.10c).

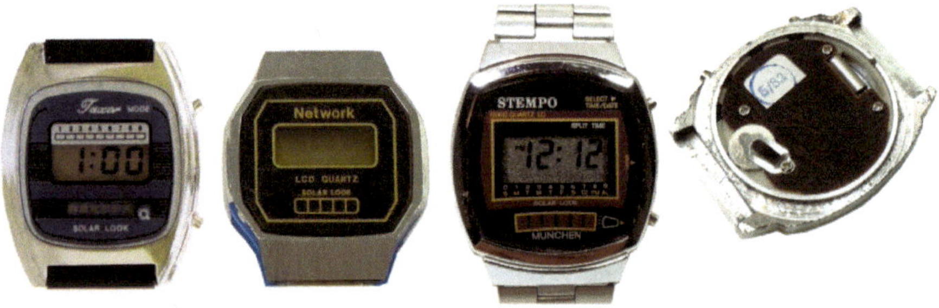

FIGURE 2.9 So-called "pseudo solar" watches from various brands. The watches prompt the "solar look" on the dial while still being powered by coin cell batteries.

(a) (b) (c)

FIGURE 2.10 Solar watches with increased functionality: (a) and (b) the Junghans radio-controlled "Mega Solar" from 1993, and (c) a Casio "batteryless" alarm chronograph.

The development of the design of solar-powered watches over time thus followed a recognizable pattern: a convergence toward more "mainstream" designs (Eggink 2017). Historic case studies of three successful technologies (the automobile, the television, and the personal computer) revealed that there is a strong correlation between the appearance of the new technology and the aesthetics of the dominant design movements. During the diffusion period of new technologies, the conformity of these technologies to the looks of the prevailing design movements of these diffusion periods increases (Snippert 2016). In other words, in the same time that these technologies became successful, they looked more and more like "normal things" (Eggink and Snippert 2017).

REFERENCES

Anonymous author (1955). Improvements in solar battery, *Journal of the Franklin Institute*, 260(1): 87–88.

Anonymous author (1978). Calculating Intentions, *New Scientist*, 79(1112): 221.

Becquerel, E. (1839). Mémoire sur les effets électriques produits sous l'influence des rayons solaires, *Comptes Rendus de L'Academie des Sciences*, 9: 561–567.

Bielefeld, C.-U. (2007). *Elektrik am Handgelenk, Geschichte und Technik der elektrischen Armbanuhr.* 1956 bis 2006. Leipzig, Germany: Engelsdorfer Verlag.

Braun (2019). Online shop: Clocks, Watches, Accessories. https://www.braun-clocks.com/calculators.html. Accessed June 26, 2019.

Chapin, D.M., Fuller, C.S., Pearson, G.L. (1954) A new silicon p-n junction photocell for converting solar radiation into electrical power, *Journal of Applied Physics*, 25: 676.

Doensen, P. (1994). *Watch, History of the Modern Wristwatch.* Gent: Snoeck Ducaju & Zoon.

Eggink, W. (2009). A chair to look to the moon: What we can learn from irrational design history for contemporary design practice, *Design Principles and Practices*, 3(1): 103–114.

Eggink, W. (2017). Sustainable technologies, Aesthetics and diffusion; a historical perspective. *Paper presented at the Design History Society Annual Conference (DHS)*, September 7–9. Norway: University of Oslo.

Eggink, W., Reinders, A.H.M.E. (2016). Product integrated PV: The future is design and styling, *Proceedings of 32nd European Photovoltaic Solar Energy Conference and Exhibition.* Munich, Germany: EU PVSEC.

Eggink, W., Snippert, J. (2017). Future Aesthetics of Technology; context specific theories from design and philosophy of technology, *The Design Journal, Design for Next: Proceedings of the 12th European Academy of Design Conference*, Rome, Italy, pp. 196–208.

Furnas, C.C. (1957). The uses of solar energy, *Solar Energy*, 1: 68–74.

Prince, M.B. (1955). Silicon solar energy converters, *Journal of Applied Physics*, 26: 68–74.

Prince, M.B. (1959). The photovoltaic solar energy converter, *Solar Energy*, 3: 35.

Reinders, A.H.M.E., van Sark, W.G.J.H.M. (2012). Product integrated photovoltaics, Chapter 34, in W.G.J.H.M. van Sark (Ed.) Photovoltaic Technology, Volume 1 in A. Sayigh (Ed.) *Comprehensive Renewable Energy*, London, UK: Elsevier.

Sillcox, L.K. (1955). Fuels of the future, *Journal of Franklin Institute*, 259(3): 183–195.

Snippert, J. (2016). Research on the influence of design on the diffusion of Technology (and how to help robots take over the world). MSc in Industrial Design Engineering, University of Twente, Enschede.

The Led Watch (2008). https://www.theledwatch.com/ledwatches.htm. Accessed May 6, 2008.

Zahl, H.A., Ziegler, H.K. (1960). Power sources for satellites and space vehicles, *Solar Energy*, 4: 32–38.

Design Case 2

Luminescent Solar Concentrator PV Designs

Angèle Reinders, Ned Ekins-Daukes, Timothy Schmidt, Hanbo Yang, Monika Michalska, Rosina Pelosi, Marcello Nitti, Lara Gillan, Dane McCamey, Elham M. Gholizadeh, Parisa Hosseinabadi, Blair Welsh, Scott Kable, and Wouter Eggink

2.1 INTRODUCTION: WHY DESIGN WITH LSC PV TECHNOLOGIES?

At present, the world's energy system is undergoing a fundamental transition. In addition to its emission reduction potential (Fthenakis and Raugei 2016), the solar energy supply is increasingly regarded as a source of economic development that creates opportunities for innovation. Most of mankind's energy demand arises from densely populated locations such as the built environment. Logically, if mankind aims to use more solar energy in the nearby future, the design of products, buildings, and cities will adapt to this new form of energy production, distribution, and consumption, leading to integrated solutions instead of mere technological instances. Luminescent solar concentrator (LSC) PV devices (see Figure DC-2.1) are particularly attractive from a design perspective since the solar collector form and color are effectively decoupled. Specifically compared to existing conventional PV modules that are dark in color and have a fixed rectangular, thin, flat shape, LSCs offer colorful features and form freedom, which significantly contributes to design opportunities that can enhance the overall functions and experience of PV applications in products and in the built environment (Reinders et al. 2018). Therefore, in this chapter two interdisciplinary

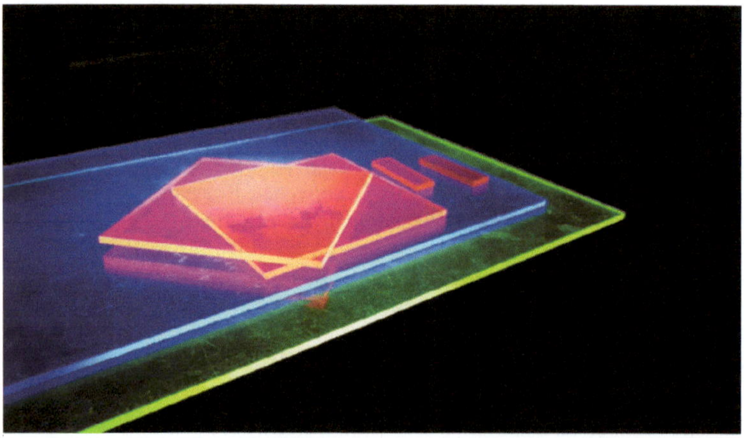

FIGURE DC-2.1 Photo of various stacked luminescent solar concentrators. (Courtesy of Center of Excellence in Exciton Science, UNSW.)

design-driven projects are presented, which cover, besides technological, also human, societal, and styling aspects. The first is a design project that was executed in 2016 at the School of Design of University of Twente (UT). The second project harnesses research advances in the Center of Excellence of Exciton Science at UNSW in Sydney, wherein the focus is on science, discovery, and the improvement of LSCs. This project took place during a one-week summer school in January 2019.

2.2 BACKGROUND REGARDING LSC PV TECHNOLOGIES

The functional principle of LSC PV was originally proposed in the late 1970s (Weber and Lambe 1976; Goetzberger 1978; Goetzberger et al. 1979, 1980). An LSC is a technology for harvesting solar energy that is comprised of a transparent, thin plate acting as a lightguide. This lightguide consists of a material with a refractive index higher than air containing luminescent pigments, called luminophores. Photons originating from solar irradiance that enter the LSC are absorbed by the luminophore and are subsequently reemitted at longer wavelengths. A large fraction of the radiation is trapped in the lightguide by total internal reflection at the material's surfaces. This is a very effective mechanism for capturing and concentrating diffuse irradiance. For an excellent overview of existing lightguide materials and luminescent dyes, we like to refer to Debije and Verbunt (2012). The primary challenge faced by LSC PV devices is increasing their photon-to-electron conversion efficiencies to improve the efficiency of the LSC device. At present, a significant amount of research is underway to develop novel luminophores, including organic dyes, such as fluorescent dyes, and inorganic dyes such as phosphors and quantum dots (QDs). Traditionally in LSC PV devices, PV technologies such as crystalline silicon or gallium arsenide solar cells are used; however, this short list is not limited to these PV technologies, and potentially, perovskite materials, amorphous silicon, and copper indium gallium deselenide (CIGS) solar cells could be applied as well. Currently, two types of characteristic LSC PV device configurations can be distinguished: LSC PV devices with cells attached to the back of the LSC and those with cells attached to the edges of the LSC. The highest efficiency achieved so far for prototypes with cells attached to the back of LSC PV modules is 5.8% (Reinders et al. 2017) and for LSC PV's with cells attached to the edges 7.1% (Slooff et al. 2008).

Apart from the use of perylenes as an organic dye in LSCs, recent developments with the application of QDs and new types of organic dyes in LSC PV devices show a far greater potential for concentration of incoming irradiance and the efficiency of these devices. Namely, Bronstein et al. (2015) showed that LSCs doped with CdSe/CdS QDs can exhibit a 30-fold concentration of absorbed irradiance. This was achieved by coupling of a wavelength-selective photonic mirror to an LSC to form an optical cavity where emitted luminescence is trapped. This type of LSC PV device contained a microsilicon solar cell. Li et al. (2016) showed that large area devices with QDs can yield good results.

They have demonstrated relatively low-loss, large-area (up to about 90 × 30 cm^2) LSCs with colloidal core/shell QDs resulting in efficiencies of more than 10% for dimensions of tens of centimeters. This is higher than the record efficiencies listed in part B, and as such this sort of breakthrough in LSC PV research is expected to result in future high-efficiency applications of LSC PV technologies.

2.3 APPROACH BY "THE POWER OF DESIGN"

The application of PV systems beyond primary electricity generation is still limited. Earlier work revealed that the design potential of PV solar cells and PV systems is often overlooked; therefore, in these projects the opportunities and challenges of designing with LSC PV devices are explored in the context of five aspects that are relevant for successful product design. These five aspects (Reinders et al. 2012) are: (1) technologies and manufacturing, (2) financial aspects, (3) societal context, (4) human factors, and (5) design and styling. In Reinders et al. (2012), these five aspects, shown in Figure DC-2.2, are discussed extensively. Here, we briefly focus on the least-known aspects, namely, "human factors" and "design and styling."

The aspect of *human factors* deals with the use of PV technologies, the perception of consumers, and their interaction with PV technologies. This is especially important in shaping a consumer's willingness to use PV technologies. For instance, from a prior study, it was concluded that respondents were quite positive about PV

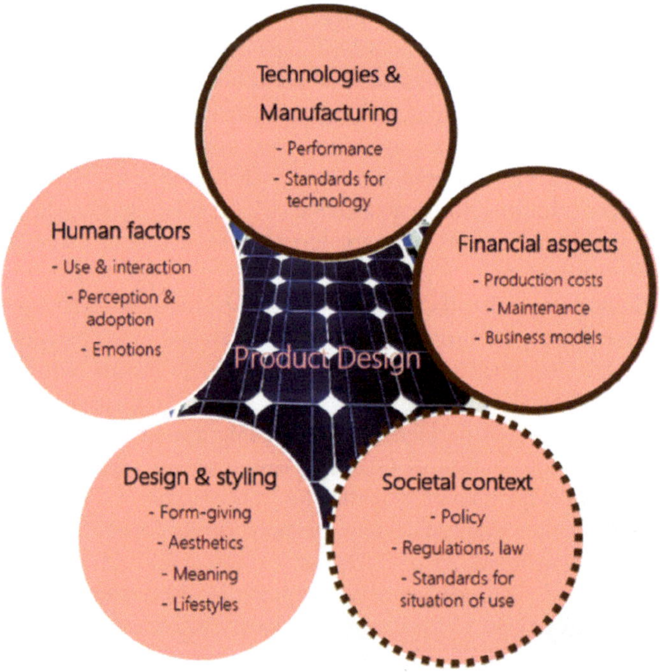

FIGURE DC-2.2 Scheme representing five aspects that are important for design with LSC PV technologies. (Courtesy of Angèle Reinders et al. 2019.)

products that have an environmentally friendly or a social character. The aspect *design and styling* deals with the appearance of PV technologies. An interesting and contemporary appearance can have a major influence on the desirability of PV-integrated products but can also play a role in making PV systems more acceptable to its users or in its environment of use. For instance, in the case of BIPV systems, well-designed objects tend to encounter less resistance and also have an increased functionality because of a positive and forgiving attitude of the user. As Donald Norman stated, "Attractive things really do work better" (Norman 2004). Moreover, the communication function of design (Crilly et al. 2004) can help to improve, in particular, the *human factors* and stimulate the acceptance of innovative technologies (Eggink and Snippert 2017).

2.4 UT'S PROJECT DESCRIPTION (2016)

A design study on possible applications for LSC PV technologies was executed within the scope of the master course "Sources of Innovation." This course is part of the Master in Industrial Design Engineering of the University of Twente. The course is especially targeted at developing novel designs for innovative technologies (Eggink and Reinders 2013). In previous years this approach was also applied to Concentrating PV and other PV technologies (Reinders 2008; Reinders et al. 2009, 2011, 2013). A series of theories about innovation processes from the book *The Power of Design* [2] is applied in an accompanied design project. In the academic year 2016–2017, about 40 students executed a design project in groups of two or three. The end results that they had to attain were a feasible concept design worked out to the level of a scaled prototype and a report that describes the design process and the application of the innovation methods.

The project resulted in a total of 16 prototyped concept designs. Due to the course's emphasis on innovation and creativity (Verganti 2009; Eggink 2011; Eggink and Rompay 2015), the student design projects were very diverse. In random order, the designed objects consisted of a tourist shelter, a garden fence, greenhouse panels, safety staircases, parking assistance in the form of trees, outdoor tables, a casing to charge cell phones, a labyrinth for parks, an illuminated bicycle frame, an LSC-powered boat, outdoor furniture, a sundial, a festival tent, and a colorful bike parking with integrated LSC PV technologies.

In the following subsections, we will highlight a subset of all the results that represent the broad application possibilities of the LSC technology on different dimensions. These dimensions are scale (from big to small), application purpose (for professional use or for leisure and entertainment), and user experience (either active or more passive). The presented designs are also illustrative examples of the use of the design features of the LSC material: colorful, transparent, relatively cheap, and bendable.

2.4.1 *AlliSee* LSC Boat

This project presents a product design for leisure activities. The design consisted of a modular platform for a small electrically powered boat for outdoor activities such as snorkeling and fishing (Figure DC-2.3).

FIGURE DC-2.3 "AlliSee" transparent boat concept by Jullian Claus, Rosan Harmens, and Hieu Nguyen.

The design makes use of one of the discerning design features of LSCs, which is the possibility to realize three-dimensional shapes. The curved form of the hull would be molded from one large sheet of LSC material. The transparency of the material adds to the user experience because the underwater world will be visible through the hull. In the design concept, the solar cells that are needed to harvest the light that is collected in the LSC hull are located in the reinforced gangway of the dinghy. Because the total LSC surface area is rather big, segmentation with an additional PV cell grid would be applied to minimize the self-absorption of the photons by the material (Figure DC-2.4).

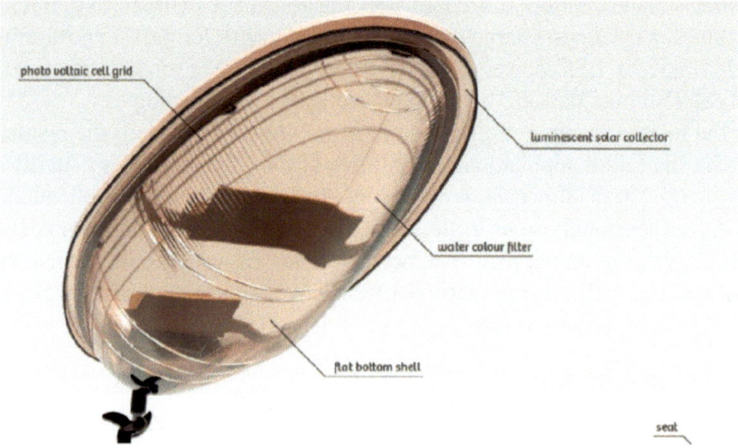

FIGURE DC-2.4 LSC transparent sculpted hull design with PV cell grid.

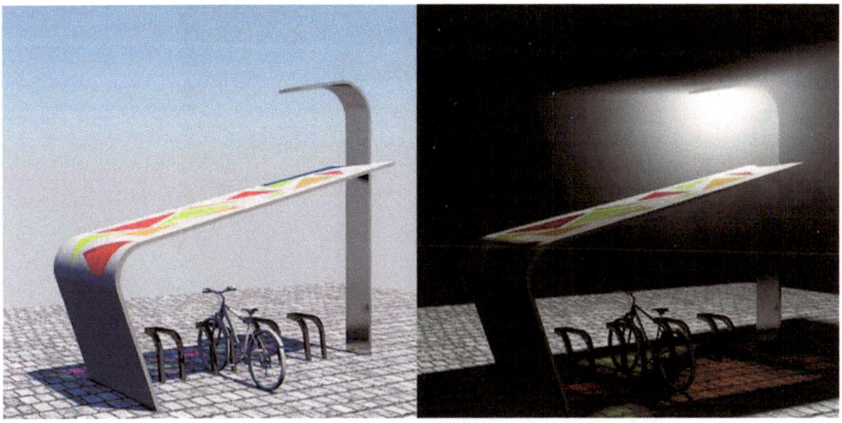

FIGURE DC-2.5 Bike parking concept with integrated LSC material and lantern by Ruben de Noord, Eduard Tudor, and Sem Vossebeld. During daytime (left) and during the night (right).

2.4.2 Bike Parking Shed with Integrated LSCs

The second example is a somewhat larger object for professional use. In this concept, LSC material is integrated in the roof of a bicycle parking (Figure DC-2.5).

The LSCs generate electrical energy that can be used to illuminate the standalone unit during the night. An integrated lantern will shed a colorful shadow pattern on the floor. This type of use of the color of the LSC material again adds to the consumer's experience in two ways. During the day it draws attention to the sunlight as an energy source, and during the night it is supposed to support the feeling of safety (Figure DC-2.6).

Apart from the safety-enhancing lighting during the night, the LSC PV panels could also be used to generate power for charging e-bikes. However, the charging station should be connected to the grid then as well, to ensure enough power supply during all seasons.

2.4.3 LSC Table for Outdoor Charging of Devices

This concept, on a smaller scale, uses the LSC material to harvest energy in an outdoor table for use at festival sites. The energy is then used to charge the cell phones of festival visitors. The cables that are provided to connect the cell phones with the table are deliberately kept short, so the visitors are unable to use their phones during the charging. This should trigger them to have real-life conversations with other visitors of the festival instead of keeping them busy with their phone (Figure DC-2.7).

The table design consists of an LSC material tabletop and four legs with a cell-phone charger integrated in each of them. The PV modules that harvest the energy are located in the leg units that surround the table surface. During the evenings, the transparent tabletop can also be lit by integrated LED strips

FIGURE DC-2.6 Colorful effect of the LSC material in the bike parking concept demonstrated in the prototype.

FIGURE DC-2.7 LSC-PV charging table concept by Niek Eggink, Gydo Nijensteen, and Mark Zwart in a typical festival environment. The charging functionality should stimulate personal contact among the visitors.

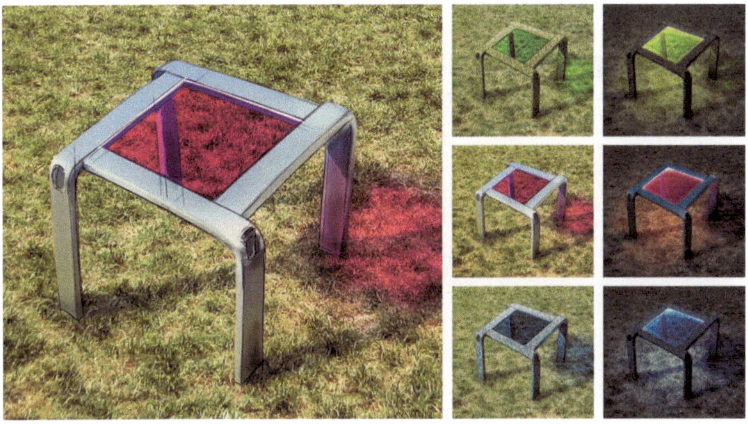

FIGURE DC-2.8 Table with LSC PV tabletop for charging cell phones. The tabletop can be lit-up by LEDs during the night (inset right).

(Figure DC-2.8). The distinct color of the LSC material and the design of the table should attract people and tickle their curiosity, in order to enhance the social function.

2.4.4 Illuminating Safety Strips

This concept consists of a modular system for illuminating outdoor surface edges, in particular stairs and steps (Figure DC-2.9).

The design is self-sufficient. With the use of LSC material in combination with solar cell strips, energy will be converted during daytime and stored by the

FIGURE DC-2.9 Application example of LSC illuminating safety strips concept by Ashley Hogt.

FIGURE DC-2.10 Impression of the construction of the modular LSC safety strips with the LSC material (magenta) surrounded by solar cells.

power controller. In the evening, the energy will be used to activate the LED module inside the casing (Figure DC-2.10).

Because one of the proposed design methods of the course is platform-driven product development [2], a lot of projects pay attention to a modular design. Also, in the case of the LED strips, several platform possibilities were explored. Apart from the more obvious options of altering the size and color of the strips, the incorporation of additional sensors to provide more functionalities was discussed. The safety strips can provide safety in a passive manner, as delineators of a "dangerous" edge, warning people to step back. However, they could also function in a more active way to create a feeling of safety by illuminating places in a playful manner, for instance, while walking public stairs when it is dark. To increase the engagement of the user, sensors will be used to make the experience interactive by responding to the users' movements (Figure DC-2.11).

FIGURE DC-2.11 Active engagement of the user by responding to the movements of people ascending and descending the staircases.

2.4.5 LSC-PV Garden Fence

This design uses the LSC material to create a multifunctional garden fence with a modular setup. The system consists of hexagonal panels with integrated light strips that can be put together to create a fence with the desired dimensions (Figure DC-2.12).

The students report that from the application of the innovation journey (one of the innovation methods provided in the course) it was shown that in previous time the LSC technology was mostly used in the public environment or applied for public utilities, so their design is an attempt to expand the use of the LSC material toward private architecture. Unlike conventional PV modules, the LSC panels can easily be cut into different shapes. The students therefore explored various possibilities with different building blocks to give the consumer the opportunity to build their own appearance (Figure DC-2.13).

FIGURE DC-2.12 Modular garden fencing system by Yanxin Wang and Biying Zhang. Scale prototype (left) and nighttime effect (right).

FIGURE DC-2.13 Exploration of basic shapes for the garden fence module with square, rhombus, pentagon, and hexagon. In the prototype, the hexagon was used.

2.4.6 Parking Lot with LSC-PV "Trees"

The parking lot design with solar-powered artificial trees is intended to support drivers with finding an empty parking spot. The tree-like structures incorporate a large LSC panel on top in the shape of a leaf that harvests the energy from the sun. The energy is used to power a system of sensors and arrows that can direct the drivers to the proper place. The trees can also be used to light the parking lot in the dark. The arrows are inserted in the tarmac and light up to show the correct direction (Figure DC-2.14).

The interactive prototype showed the functionality of the system (Figure DC-2.15). The shapes of the LSC-PV energy sources are resembling a tree to add to the idea of sustainability of solar power.

2.4.7 LSC-PV Labyrinth

The labyrinth that is built from large LSC-PV panels is one of the more explicit examples of the benefits of the relative cheapness of the material. The labyrinth is meant for public parks and can easily be modified, enlarged, or altered to stay interesting for the public (Figure DC-2.16).

The design concept uses the energy that is produced by the solar panels to make the labyrinth interactive. A number of rotating panels will dynamically change the route through the labyrinth in reaction to the people inside. The changes can either direct the user to the exit or conversely make the route more difficult.

2.4.8 LSC-Sustainable Festival Tent

The largest concept in terms of scale is the sustainable festival tent. This concept uses LSC PV modules to generate a part of its own electricity consumption. In terms of design features of the LSC material, the color of the luminescent dyes in the LSC panels is used to create a special atmosphere

FIGURE DC-2.14 LSC-PV "trees" concept for an interactive parking lot, by Samantha IJkhout and Jennifer van Zeil.

FIGURE DC-2.15 Prototype of the interactive parking lot. If a parking spot is occupied (mimicked by the hand on the left), drivers are directed to the other spots by the lighted arrows.

FIGURE DC-2.16 LSC-PV Labyrinth by Natasha Tanushevska, Renée Schrauwen, and Eline de Wall.

inside the tent. Because most festivals are held both during day and night, the lighting effect is both special when the sunlight comes through the surface and when the tent is lit with artificial lighting (Figure DC-2.17). The students made a working prototype at scale to demonstrate the lighting effect (Figure DC-2.18).

The tent surface consists of triangular modules. When the night falls, the LSC panels are lit with the use of LED strips that are placed at one edge of the triangles. At the other two edges, solar cells are attached to catch the energy of the light that is trapped (Figure DC-2.19).

FIGURE DC-2.17 Impression of the sustainable festival tent by Rolf van der Toom, Steven Oonk, and Kasper Schriek.

FIGURE DC-2.18 Scale model of the sustainable festival tent, demonstrating the lighting effect.

The students calculated that the tent could not provide enough energy when powered by LSC-PV panels only and therefore suggested to incorporate both LSC-PV modules and conventional PV modules in a complete festival setup.

2.4.9 LSC-PV Urban Street Furniture

In this project, the students designed multifunctional street furniture (Figure DC-2.20). The playful modules of the furniture contain LSC panels that

FIGURE DC-2.19 Impression of the construction of the modular LSC tent segments, with both LSC and conventional PV panels. On the bottom right is the setup of an LSC PV panel with solar cells and an LED strip around the edges.

FIGURE DC-2.20 Street furniture concept by Canxuan Li, Théo Sauzon, and Jules Troia.

provide energy for integrated cell-phone chargers, similar to the festival table described earlier. The colorful units represent the different block shapes of the popular Tetris game and can be combined in several ways. This invites the users to build their own "urban landscape." To enhance the association with the game, the students named their system "Letris." Because the blocks are meant to be positioned to the users wish, they contain LSC panels at each side of the cubes. The panels also incorporate a grid to minimize the internal loss through self-absorption of the photons by the material (Figure DC-2.21).

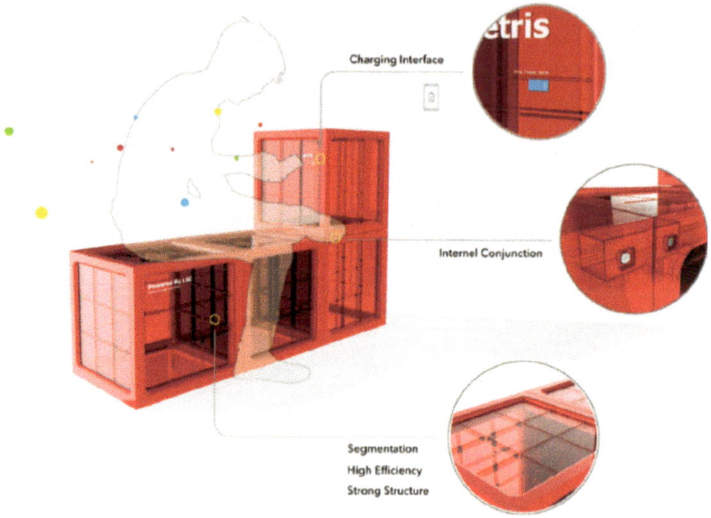

FIGURE DC-2.21 Details of the Letris urban furniture concept with charging interface, connections, and LSC grid surfaces.

2.5 UNSW'S PROJECT DESCRIPTION (2019)

To further explore the design potential of LSC PV applications, a summer school was organized on the topic of "Exciton Solar + Design" in January 2019. This summer school, which was hosted by the ARC Centre of Excellence in Exciton Science and the School of Photovoltaic & Renewable Engineering at UNSW Sydney, aimed to engage both scientists and designers and included formal lectures and hands-on sessions. Topics covered at this five-day summer school were: Excitonics and Solar Cells, Next-Generation Photovoltaics, Luminescent Solar Concentrators, Photovoltaic Design and Product Development, Building-Integrated PV Production, and Design and Practical Measurements on LSC PV devices.

The summer school was attended by 15 participants with backgrounds in science, ranging from chemistry, physics, and electronic engineering. Besides better insights of the field of LSC PV, the summer school resulted in several conceptual designs from the participants. In this section, we will present the most prominent conceptual design projects with a short explanation by their creators who are both scientists and designers.

2.5.1 Solar Lilies by Hanbo Yang and Monika Michalska

Aesthetics and perceived versatility gave birth to Solar Lilies. This product concept reimagines LSCs into petal shapes that fit onto a replaceable base where solar cells are attached (Figure DC-2.22). The changeable base allows the lilies to decorate various environments from a garden pool to trees. Furthermore, the lilies, when connected in a series, can charge air pods or any small electronic device in cloudy weather or indoors, which provides an attractive complement to traditional solar solutions that require direct sunlight.

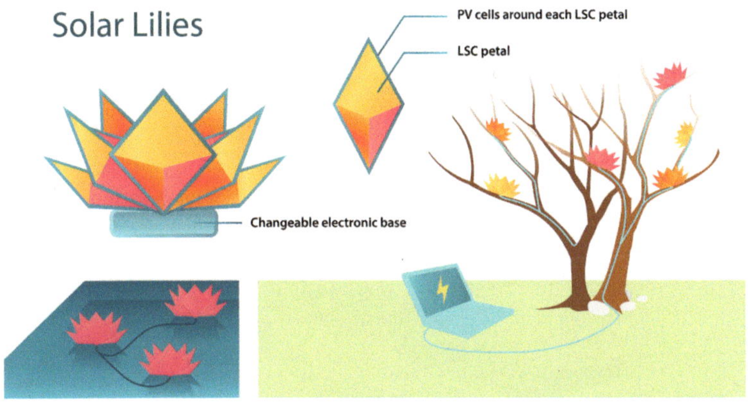

FIGURE DC-2.22 Visual representation of the Solar Lilies design project. (Courtesy of Hanbo Yang and Monika Michalska.)

The colors of the petals are entirely self-configurable in the color range of yellowish to reddish colors. Fluorescent dyes that cover different regions of the solar spectrum are designed to generate higher-power outputs. The established polymer host materials are also extremely lightweight and can be manufactured in flat or bent shapes to create various geometries for the design of a solar lily.

Figure 4 shows a solar lily in a Computer Aided Design (CAD) model. Each of the eight petals made of poly methyl methacrylate (PMMA) doped with 150 ppm of Lumogen Red 305 dye consists of a pentagon with a thickness of 1 cm and length of 8.5 cm. The entire Solar Lily spans across 21 cm in length. Solar cells ($W = 0.03$ cm, $L = 3$ cm, and $H = 1.1$ cm) are attached to the petal edges, which are connected to the purple base, which has a diameter of 5 cm. The other four edges of the petal are covered with mirrors to increase total internal reflections and to reduce reabsorption losses.

To estimate the performance of a solar lily, ray-tracing simulations have been executed for the design shown in Figure DC-2.4 using the advanced software package LightTools (Synopsis 2019). Under direct AM 1.5G solar irradiance of 1000 W/m^2 with a diffuse share of 200 W/m^2, the average solar irradiance power collected by the surface of each top petal is 0.021 W with an error estimate of 4%, whereas this value is 0.036 W for a bottom petal. To further improve the power output, a higher concentration of luminophores can be used. The geometry of the Solar Lily will also affect the amount of power collected as such the initial design shown in Figure DC-2.23 can be further optimized.

2.5.2 Arctic Fire Bottle by Lara Gillan, Dane McCamey, Timothy Schmidt, Elham M. Gholizadeh, Parisa Hosseinabadi, Rosina Pelosi, Blair Welsh, and Scott Kable

This LSC innovation in a bottle, called Arctic Fire Bottle, as shown in Figure 2.24, is designed to heat and cool liquids without a plug-in. From steaming hot chocolate on Mt. Kosciuszko to a chilled beer at Bondi, you won't need a

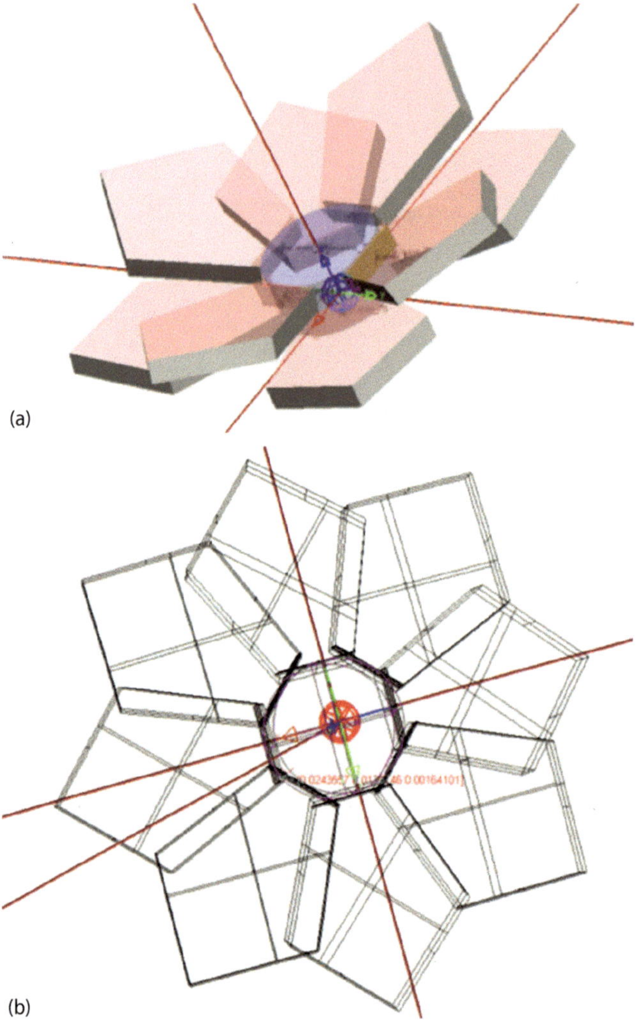

FIGURE DC-2.23 CAD presentation of a solar lily. On top (a), side view, and at the bottom (b), transparent view from above. (Courtesy of Hanbo Yang.)

campfire or a cold-box, known in Australia as an esky. The temperature control on top of this bottle is set externally, so you can sit back and relax while Arctic Fire gets to work.

The LSC body absorbs direct and diffuse light and funnels it to the attached solar cell. The direct current (DC) power production is then regulated and stored in a battery (Figure DC-2.25). On selecting the heating function, the induction heating plate draws on the batteries' power to warm liquids. Cooling is achieved by thermodynamic adiabatic cooling. By selecting the cooling function, compressed gas is released, thereby rapidly cooling the surrounding chamber and any liquid contained within it.

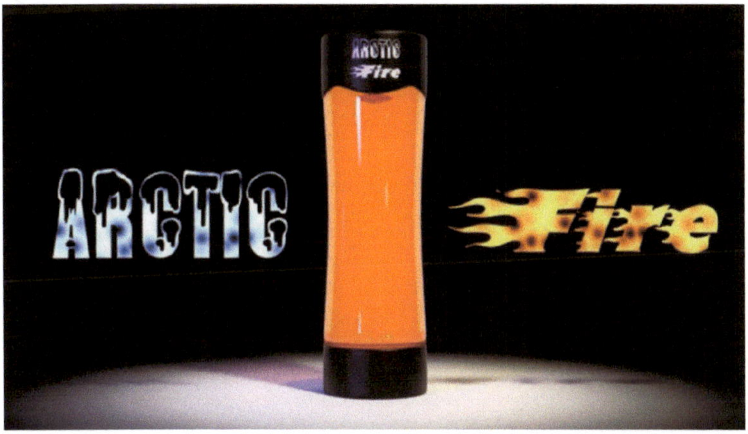

FIGURE DC-2.24 Visual representation of the Arctic Fire bottle design project. (Courtesy of Blair Welsh.)

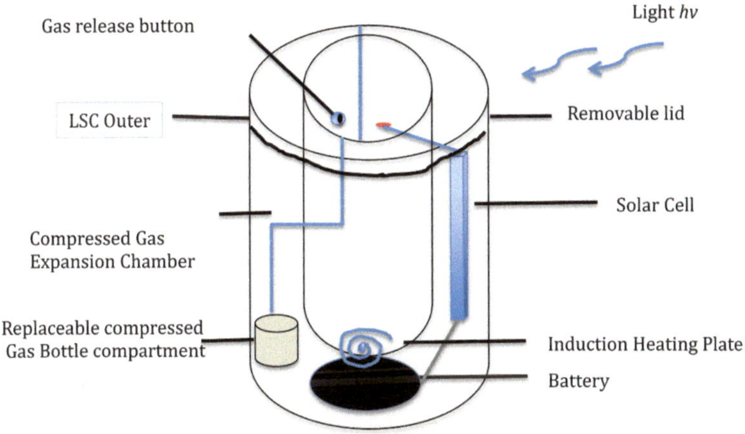

FIGURE DC-2.25 Schematic of the heating and cooling bottle shown above.

2.5.3 Colorful, Customizable Building Elements by Ned Ekins-Daukes, Marcello Nitti, and Angèle Reinders

Bifacial solar cells offer some additional design flexibility since they permit illumination on both sides. Therefore, the first LSC PV device with bifacial solar cells is presented in Figure DC-2.26. Each unit cell in this device, see Figure 2.8, consists of a squared lightguide that is covered at each side by four solar cells, which are surrounded by other cubes, which are also acting as lightguides.

In this example, each cube measures 1 × 1 × 1 cm; however, the optimal sizing would vary depending on the luminophore and host material used. In this conceptual design, the dispersed luminophore in the PMMA lightguide matrices

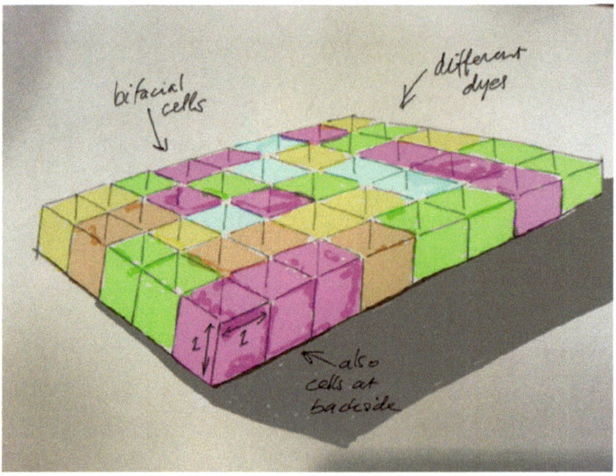

FIGURE DC-2.26 A conceptual design of a PV building element with LSC bifacial PV technologies showing the customizability features of color. (Courtesy of Reinders and Ekins-Daukes.)

is the organic Lumogen Orange 240 dye. In practice, the choice of different dyes can lead to customizable color schemes for this sort of new LSC bifacial PV modules, which would make them, in particular, suitable for BIPV applications.

To estimate the performance of this design of a bifacial PV module, ray-tracing simulations have been executed of the unit cell shown in Figure DC-2.27 using the advanced software package LightTools (Synopsis 2019). In the simulation, the total irradiance was set at 1000 W/m^2 on top of the middle lightguide surface.

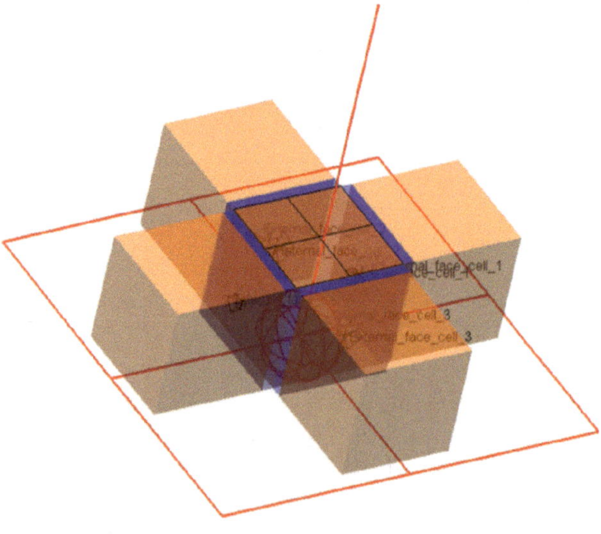

FIGURE DC-2.27 CAD presentation of a unit cell in the LSC PV module shown in Figure DC-2.26. (Courtesy of Marcello Nitti.)

The irradiance is divided in 70% (700 W/m^2) direct irradiance and 30% (300 W/m^2) diffuse irradiance. The direct source is located front facing the device far away from the device itself, thus acting as the Sun; while the diffuse irradiance source is surrounding the whole device. The ray-tracing simulations have been performed while changing the dye concentration in the PMMA lightguide.

These simulation results are depicted in Figures DC-2.28 and DC-2.29, where internally faced surface and externally faced surfaces of cells are separately

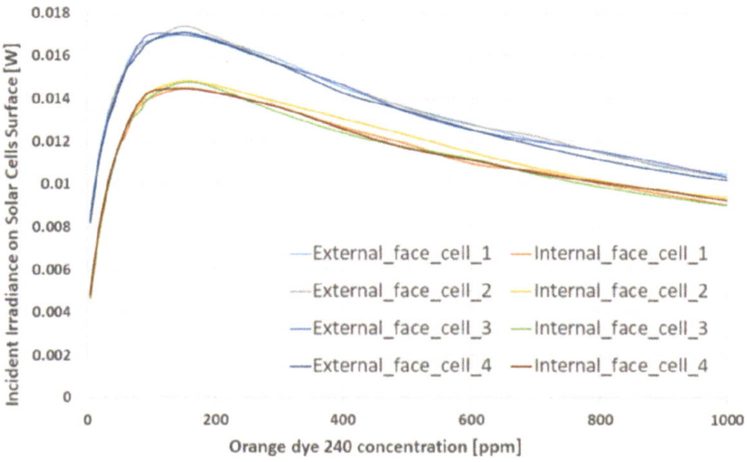

FIGURE DC-2.28 Incident irradiance on solar cell surfaces for a unit cell of a bifacial LSC PV module under 1000 W/m^2 irradiance, see Figure DC-2.27, versus the concentration of Lumogen Orange 240 dye.

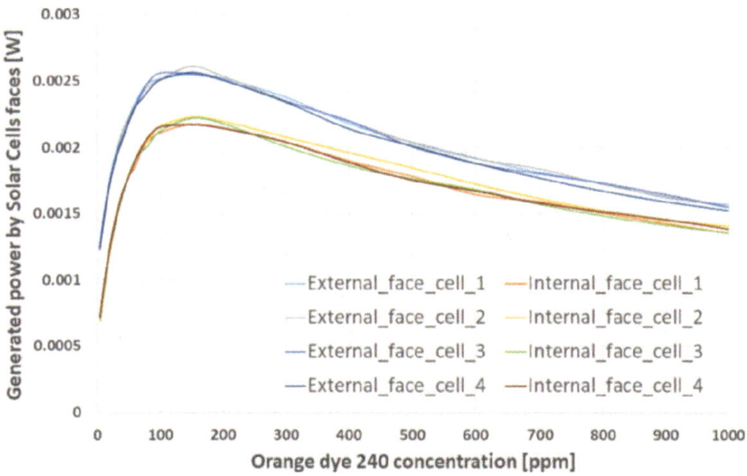

FIGURE DC-2.29 Generated power by solar cells of a unit cell of a bifacial LSC PV module at 15% cell efficiency under 1000 W/m^2, see Figures DC-2.27 and DC-2.28, versus the concentration of Lumogen Orange 240 dye.

represented. As can be seen, the incident irradiance power on the solar cell surfaces (of 1 × 1 cm) increases with increasing dye concentration, while it reaches a maximum value and then starts to decrease due to the occurrence of, among others, reabsorption losses.

The total incident irradiance power on the top surface of each cube is 0.082 W. In this situation, the maximum average irradiance is 0.015 W on each internally directed surface of the solar cells at a dye concentration of 150 ppm; while the externally directed surfaces receive a maximum irradiance of 0.017 W due to greater exposure to light. Subsequently, when assuming a modest cell efficiency of 15%, the externally directed solar cells will yield a maximum power of 0.0026 W per cell, and the internal cell surfaces will produce 0.0022 W per cell. A simple calculation of the effect of interconnecting many unit cells in a PV module in which all solar PV cells will be internally directed results in a performance estimation of 0.009 W per 1 cm^2, which is similar to 90 W/m^2 under irradiance of 1000 W/m^2 and cells with an efficiency of 15%. As such, by this design the performance of the LSC PV technology not only looks aesthetically pleasing but will also generate a competitive efficiency of 9%. By optimizing the cell efficiency, the geometry of the unit cells and by adding semitransparent mirrors this initial efficiency could be increased. Assuming cell efficiencies of above 20%, the module efficiency of this sort of transparent LSC bifacial PV elements could easily exceed 10%.

2.5.4 Other Designs Using LSC PV Technologies

In addition to the three examples shown in this paper, the design project resulted in several other innovative conceptual designs: intelligent safety helmets, colorful poultry housing, and a very useful design of a self-watering plant pot.

2.5.5 Outdoor Experiments with LSC PV Samples

Following the conclusion of the design phase, the summer-school participants spent a day at Coogee Beach evaluating the performance of LSC devices under real outdoor conditions around noon with high irradiance and high temperature. Using only a standard multimeter and a basic LSC device consisting of silicon cells and dye-doped PMMA (Figure DC-2.30), they estimated the absorbed solar flux in the dye-doped PMMA sheet and made an estimate for the flux gain achieved at the edge of the sheet.

The aim of the exercise was to help students understand the often-confused measures for LSC performance. The geometrical gain G is the ratio of the surface area of the collector to the surface area of solar cells employed and is a principle figure of merit for conventional concentrator systems. The students could readily calculate the geometrical gain simply by measuring the relevant dimensions of their PMMA sheet and the PV cell supplied. The flux gain F is the ratio of the measured short-circuit current density (J_{sc}) of the PV cell when integrated into an LSC and the PV cell short-circuit current when illuminated directly without any LSC. The ratio of F/G provides a useful figure of merit for a solar concentrator. If G provides the potential for solar concentration, F provides the extent to which concentration is actually achieved over the spectral range of

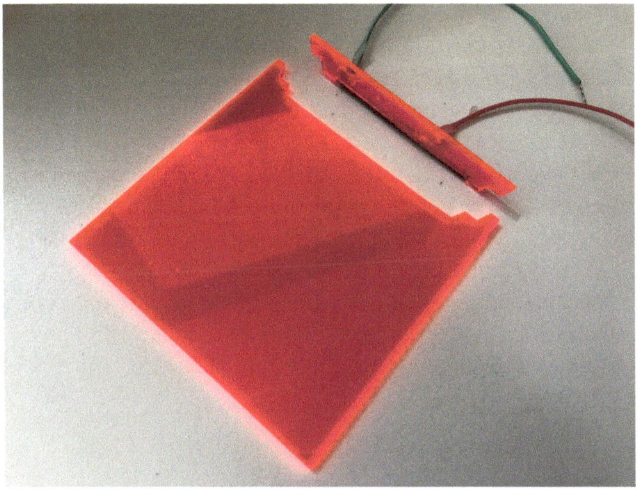

FIGURE DC-2.30 Photograph of the LSC collector and the PV cell used in the experiments. The cell was premounted on a piece of PMMA to aid the construction of the LSC during the practical session.

the PV cell. Since luminescent solar concentrators generally do not absorb the same bandwidth of sunlight as the PV cell, there is a further figure of merit: the absorbed flux gain E, which is the ratio of the J_{sc} of the solar cell in the LSC and the absorbed flux in the LSC. The students could determine this by measuring the J_{sc} above and below the LSC material. These figures of merit can then be used to calculate the optical efficiency of the system $\Phi_{opt} = E/G$ that indicates the efficiency that light absorbed by the LSC is directed to the PV cell. For the purpose of overall solar-power generation, the flux efficiency of the system is more useful $\eta_{\varphi} = F/G$, since this figure of merit compares how efficiently light that could be absorbed by the PV cell is guided through the LSC. To date, LSCs typically have F/G values <0.4 indicating that there is considerable room for improvement in the flux efficiency η_{φ} of the system.

Students were given an LSC, a PV cell (as shown in Figure 2.30), a multimeter, some UV-cured epoxy glue, masking tape, electrical tape, aluminum foil tape, and provided with a worksheet that we reproduced below with typical values obtained by the students.

1. Measure the dimensions of your solar cell. Typical values: 6 × 0.3 cm
2. Measure the dimensions of your LSC. Typical values: 10 × 10 × 0.3 cm
3. Measure your cell's I_{sc} in full sunlight. The result will fluctuate, so record the maximum value you see. This result is A *Unfiltered Current*. Typical value: 75 mA
4. Measure the cell's I_{sc} filtered by the collector, that is, place the solar cell under the LSC and measure the current. Again, measure the maximum value. This value is B *Filtered Current*. Typical value: 55 mA
5. The absorbed current in the LSC is $C = (A-B)$. Typical value: 20 mA

6. Fit your solar cell into the notch cut in the collector. Observe if there is a large air gap. If so, you may need to use a filler strip.
   a. If a filler strip is required, glue this first to the solar cell using the UV-cured epoxy glue. Be careful not to get this onto your hands! Use the applicator.
   b. Glue the solar cell into the notch using the UV-cured epoxy glue. Be careful not to get this onto your hands! Use the applicator.
   c. Be careful to make these joins as optically transparent as possible. Photon scattering here is a loss!
7. Place black electrical tape across the join between the solar cell and the collector to prevent spurious excitation of the solar cell.
8. Measure the LSC device's I_{sc}. This value is *D, LSC Current*. Typical value: 76 mA

Analysis:

1. The absorbed flux gain is $E = D/C$. Typical value: 3.8
2. The flux gain is $F = D/A$. Typical value: 1.0
3. Calculate the geometric gain, G. This is the ratio of the collection area to the area of the solar cell(s).
   Typical value: 47
4. Calculate the optical efficiency, $\Phi_{opt} = E/G$. Typical value: 8%
5. Calculate the flux efficiency, $\eta_{\varphi} = F/G$. Typical value: $\eta_{\varphi} = 2\%$

Summarizing, students typically found that their dye-doped PMMA achieved a geometrical flux gain over the absorbed photon flux in the dye-doped PMMA sheet but not an overall energy gain. In agreement with almost all published results so far on large-area LSC PV devices with cells attached to the edges, a photon flux gain relative to the LSC absorbed solar flux is commonly achieved, yet an overall, broadband optical energy concentration remains elusive. Nevertheless, the enthusiastic participants chained a few of their collectors together to successfully power a small electromotor.

2.6 DISCUSSION AND CONCLUSIONS

The LSC PV is an alluring technology that affords a freedom in design that is not easily achieved with other PV devices. Therefore, it can play a major role in creating PV applications, which will be easily accepted by many users. By following a formal design process offered by the methods of "The Power of Design" (Reinders et al. 2012), the participants of both design projects at UT and UNSW generated some ambitious ideas for attractive, solar-powered products, some of them showing a new form of aesthetic. In the first project, the focus was clearly on design only, while in the second project, the "Exciton Solar + Design" summer school, the students' interests were either on aesthetics, such as by the beautiful design of the Solar Lilies, or on an enhanced efficiency, such as the bifacial LSC PV module. This project was supported by experimental field studies and by simulations to verify the expected performance of the new LSC PV designs. In many cases, the power density of sunlight would be sufficient to generate the necessary energy, but the present performance of LSC technology should be improved in the nearby future to realize the more diverse applications. Therefore, LSC research and development underway at the ARC Centre for Exciton Science will provide the necessary technology push. Also, the use of simulation tools for ray-tracing appeared to be useful for evaluating the theoretical optical performance of conceptual LSC PV designs and for design optimizations under conditions of use. On the basis of the simulations it seems well possible that LSC PV devices can achieve efficiencies above 10% (Aghaei et al. 2020) and hence become a serious alternative for existing PV technologies while at the same time creating an immense design potential.

ACKNOWLEDGMENTS

Hanbo Yang thanks the ARC for support through grant CE170100026.

REFERENCES

Aghaei, M., Nitti, M., Ekins-Daukes, N. and Reinders, A. (2020). Simulation of a novel configuration for luminescent solar concentrator photovoltaic devices using bifacial silicon solar cells, *Applied Sciences*, 10, 871. doi:10.3390/app10030871.

Bronstein, N.D. et al. (2015). quantum dot luminescent concentrator cavity exhibiting 30-fold concentration, *ACS Photonics*, 2(11): 1576–1583.

Crilly, N. et al. (2004). Seeing things: Consumer response to the visual domain in product design, *Design Studies*, 25(6): 547–577.

Debije, M.G., Verbunt, P.P.C. (2012). Thirty years of luminescent solar concentrator research: Solar energy for the built environment, *Advanced Energy Materials*, 2(1): 12–35.

Eggink, W. (2011). Disruptive Images: stimulating creative solutions by visualizing the design vision. In: Kovacevic, A. Ion, A., McMahon, Buck, C. and Hogarth, P. (Eds.), *Proceedings of the 13th International Conference on Engineering and Product Design Education*, The Design Society, pp. 97–102.

Eggink, W., Reinders, A.H.M.E. (2013). Explaining the design & styling of future products. In: Bohemia, E., Ion, B., Kovacevic, A. et al. (Eds.), *Proceedings of the 15th International Conference on Engineering and Product Design Education*. Dublin: The Design Society.

Eggink, W., Snippert, J. (2017). Future Aesthetics of Technology; context specific theories from design and philosophy of technology, *The Design Journal*, 12: 196–208.

Eggink, W., van Rompay, T.J.L. (2015). Shake it Off; radical meaning innovation in Product Design. *Proceedings of the 5th Creativity and Innovation Management (CIM)*, September 1–2. Enschede (Netherlands).

Fthenakis, V., Raugei, M. (2016). Environmental life-cycle assessment of photovoltaic systems, *The Performance of Photovoltaic (PV) Systems: Modelling, Measurement and Assessment*, pp. 209–232.

Goetzberger, A. (1978). Fluorescent solar energy collectors: Operating conditions with diffuse light, *Applied Physics*, 16(4): 399–404.

Goetzberger, A. et al. (1979). Solar energy conversion with fluorescent collectors, *Elektronikpraxis*, 14(3): 31–36.

Goetzberger, A. et al. (1980). Principles of solar energy conversion on the basis of fluorescent collectors, *Electric Power Research Institute (Report)*, 2: 1175–1184.

Li, H. et al. (2016). Doctor-blade deposition of quantum dots onto standard window glass for low-loss large-area luminescent solar concentrators, *Nature Energy*, 1(12): 16157.

Norman, D.A. (2004). *Emotional Design: Why We Love (or Hate) Everyday Things New York*. New York: Basic Books.

Reinders, A.H.M.E. (2008). Product-integrated PV applications–How industrial design methods yield innovative PV-powered products, *Proceedings of 33th IEEE-PVSC*. San Diego, CA.

Reinders, A.H.M.E., de Boer, A. (2011). Product integrated PV and innovative product design, *Proceedings of 28th EU PVSEC*. Hamburg, Germany.

Reinders, A.H.M.E., de Boer, A., de Winter, A., Haverlag, M. (2009). Designing PV powered LED products–Sensing new opportunities for advanced technologies, *Proceedings of 34th IEEE-PVSC*. Philadelphia, PA.

Reinders, A.H.M.E., de Borja, J., de Boer, A. (2011). Product-integrated sustainable energy technologies–Six years of experience with innovation and sustainability, *Proceedings of IASDR*. Delft, the Netherlands: Diversity and Unity.

Reinders, A.H.M.E., Diehl, J.C., Brezet, J.C. (2012). *The Power of Design: Product Innovation in Sustainable Energy Technologies*. Chichester, UK: John Wiley & Sons, by hard-cover book, London: John Wiley & Sons.

Reinders, A.H.M.E., Ekins-Daukes, N., Schmidt, T., Yang, H., Michalska, M., Pelosi, R., Nitti, M., Gillan, L., McCamey, D., Gholizadeh, E.M., Hosseinabadi, P., Welsh, B., Kable, S. (2019). Designing with luminescent solar concentrator photovoltaics, *Proceedings of 46th IEEE-PVSC*, Chicago, IL.

Reinders, A.H.M.E., Kishore, R., Slooff, L., Eggink, W. (2018). Luminescent solar concentrator photovoltaic designs, *Japanese Journal of Applied Physics*, 57(8): 08RD10.

Reinders, A.H.M.E., Wiesenfarth, M., King, R.R. (2013). Conceptual product development with integrated concentrating PV systems–CPV in the built environment from a designer's perspective, *Proceedings of 39th IEEE-PVSC*. Tampa, FL.

Reinders, A.H.M.E. et al. (2017). Measured efficiency of a luminescent solar concentrator PV module called leaf roof, *IEEE Journal of Photovoltaics*, 7(6): 1663–1666.

Slooff, L.H., Bende, E.E., Burgers, A.R., Budel T., Pravettoni, M., Kenny, R.P., Dunlop, E.D., Büchtemann, A. (2008). A luminescent solar concentrator with 7.1% power conversion efficiency, *Physica Status Solidi*, 2(6): 257–259.

Synopsis, LightTools, software https://www.synopsys.com/optical-solutions/lighttools.html, last visited on June 2, 2019.

Verganti, R. (2009). *Design-Driven Innovation: Changing the Rules of Competition by Radically Innovating What Things Mean*. Boston, MA: Harvard Business Press.

Weber, W.H., Lambe, J. (1976). Luminescent greenhouse collector for solar radiation, *Applied Optics*, 15(10): 2299–2300.

3 Photovoltaic Technologies in the Context of Design

Angèle Reinders and Georgia Apostolou

CONTENTS

3.1 A BRIEF OVERVIEW OF PHOTOVOLTAIC TECHNOLOGIES

In this section, a brief overview is given of photovoltaic (PV) materials, PV cells, and PV modules as well as several standard indicators, which are common for addressing the performance of PV applications. Detailed information about the physics principles, materials, and production processes of PV technologies can be found in the recently published book *Photovoltaic Solar Energy—From Fundamentals to Applications* (Reinders et al. 2017). This book serves as a reference for this chapter, which shows merely the tip of the iceberg of the numerous technological developments in the field of photovoltaics. Namely, PV cells are made from several semiconductor materials using different technologies and different device structures. They have one operational functional aspect in common: the photovoltaic effect, which is the creation of voltage and electric current in a material upon exposure to light.

The most well-known solar cell technology is the single crystal silicon wafer-based solar cell, indicated by c-Si. It has improved significantly in the past nearly 70 years, and it has always been the dominant solar cell technology (FhG-ISE 2019).

Crystalline silicon solar cell technology represents also multicrystalline silicon solar cells, indicated by m-Si. Both c-Si and m-Si are so-called first-generation solar cells (Green 2003).

Besides this, thin-film solar cells have been developed with the intention to find a cheaper alternative for crystalline solar cell technology by using less material. These are called second-generation solar cells. Several semiconductor materials allow for the production of thin films, namely copper indium gallium diselenide ($CuInGaSe_2$), abbreviated to CIGS; cadmium telluride (CdTe); hydrogenated amorphous silicon (a-Si:H); and thin-film polycrystalline silicon (p-Si). Also, third-generation PV cells exist, which can be made from organic materials: in the first place, dye-sensitized solar cells (DSSC) consisting of titanium oxide nanocrystals covered with light-absorbing organic molecules. Second, polymer organic solar cells can be made from conducting polymers with light-absorbing C_{60} molecules. In the recent years, a new type of solar cell has emerged made of perovskite materials which, in 2019, were under development in many R&D labs and not widely commercially available yet.

Solar cells can also be made from compounds from the elements Ga, As, In, P, and Al, also denoted as III–V PV technologies. The specific cells are named after their compounds, for instance, GaAs, GaInP, or InP. These PV cells have been developed for space applications because of their high efficiency. They can also be applied in terrestrial concentrator systems. By stacking layers of different III–V compounds, tandem cells and multiple junction (MJ) cells can be made up to six or even more junctions, resulting in very high efficiencies.

Recent developments in solar cells and their applications happen in various directions, and most prominent ones are heterojunction solar cells (which have a combined crystalline and amorphous silicon device structure, e.g., the so-called HIT cells); new hybrid tandem structures using different semiconductor materials (e.g., silicon perovskite tandem cells); and bifacial silicon solar cells, which can generate power at both sides of a cell.

3.2 EFFICIENCY OF SOLAR CELLS

The sensitivity of each solar cell strongly depends on the wavelength of the light falling onto the solar cell. The sensitivity as a function of wavelength is called the spectral response (expressed in [A/m^2]/[W/m^2] or in [A/W]). This spectral response strongly influences the external efficiency of a PV solar cell, which is an important variable in the design of products that contain integrated PV. The efficiency namely determines the power that can be produced for the application designed. It depends on the PV material and technology of the cell and the intensity of irradiance that falls on a PV cell surface. Efficiency, η (in %) is given by:

$$\eta = P / A \cdot G \tag{3.1}$$

where P is the power produced by the cell (in W), A the area of the solar cell area (in m^2), and G the global irradiance (in W/m^2). Under standard test conditions of G = 1000 W/m^2, a standard spectrum called AM1.5 and a cell temperature of T_{cell} = 25°C; this efficiency is the STC efficiency, denoted as η_{STC} (in %). Figure 3.1

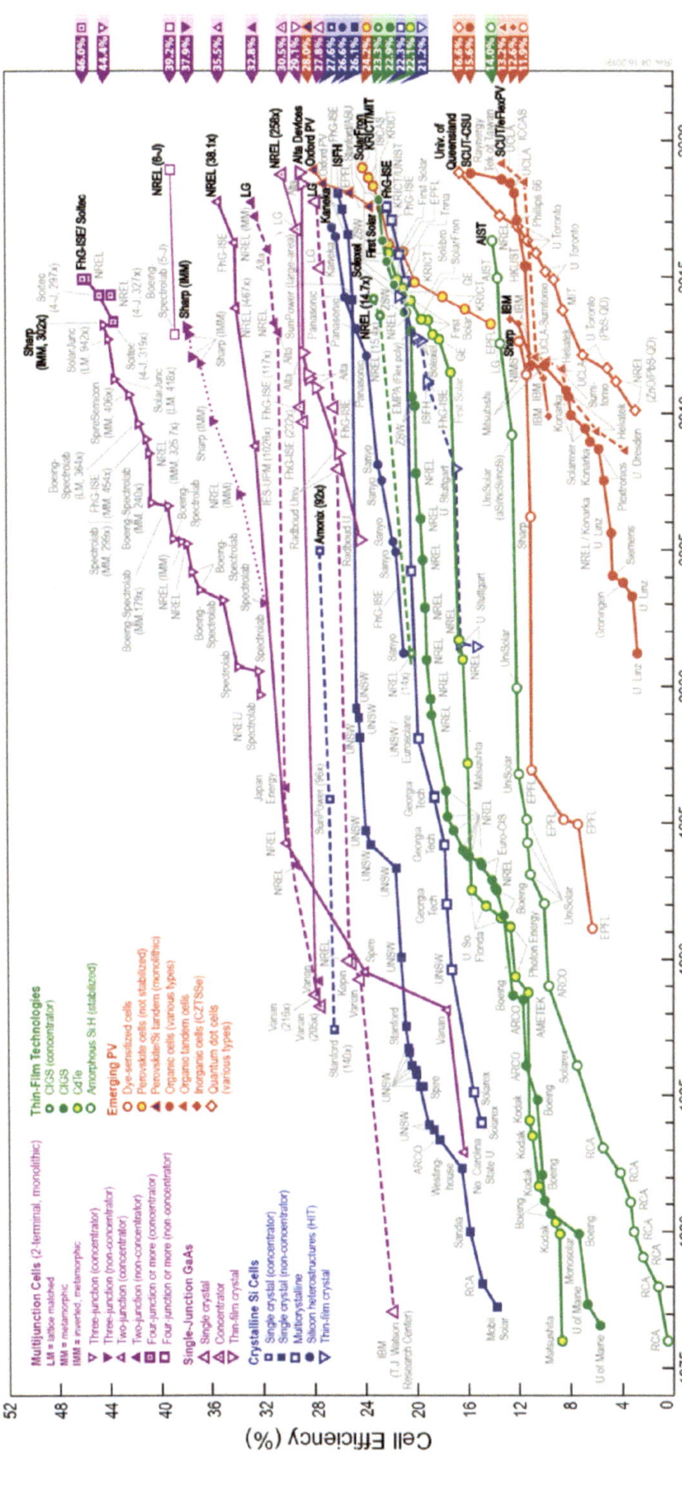

FIGURE 3.1 Best research-cell efficiency chart for several PV technologies. (NREL, National Renewable Energy Laboratories, Best Research-Cell Efficiency Chart, https://www.nrel.gov/pv/cell-efficiency.html, Last visited July 2, 2019.)

shows maximum efficiencies for the PV technologies mentioned in Section 3.1, which are developed and optimized in R&D labs. As can be seen, efficiencies range from about 12% for thin-film amorphous silicon cells to close to 40% for multijunction solar cells made of III–V compounds. Solar cells under concentrated irradiance efficiencies of close to 50% have been achieved. On the contrary, most commercially available mainstream silicon PV modules have an η_{STC} in the range of 16% to 22%. Therefore, for highly efficient PV applications, it might be interesting to explore other opportunities in the best research-cell efficiency chart shown in Figure 3.1.

The DC power produced by a PV cell, PV module, or PV system under STC conditions, P_{nom}, is named nominal power and given in the unit Wattpeak (abbreviated as Wp).

In addition, the temperature of the PV cell and the spectral distribution of the light affect the efficiency. This can be explained by the physical functioning of solar cells, which, under illumination, generate a photocurrent. This photocurrent is in most cases linearly related to the intensity of irradiance. Because of the semiconductor materials in the PV cell, the electric behavior of a PV cell can be represented by a current source in parallel with two diodes, D_1 and D_2. A series resistance, R_s, and a parallel resistance, R_{sh}, add to this electric circuit as shown schematically in Figure 3.2.

The electric behavior or current-voltage characteristic (I–V curve) of a PV cell is described by Green (1982):

$$I = I_{ph} - I_{s1}\left(e^{\frac{q(V+IR_s)}{n_1 kT}} - 1\right) - I_{s2}\left(e^{\frac{q(V+IR_s)}{n_2 kT}} - 1\right) - \frac{V+IR_s}{R_{sh}}. \tag{3.2}$$

Here I_{s1} and I_{s2} (in A) are the saturation currents of the two diodes, and n_1 and n_2 are the quality factors of these two diodes. In general, n_1 will not deviate much from 1, and usually $n_2 = 2$ if no imperfections occur. V (in V) is the voltage over the circuit, T (in K) is the temperature, and k is the Boltzmann constant, which is 1.38064852 10^{-23} J/K. Figure 3.3 shows the I–V curve of a solar cell at a certain irradiance and temperature. It crosses the y-axis at the open-circuit voltage, V_{oc} and the x-axis at the short-circuit current, I_{sc}. The shape of the I–V curve changes with irradiance and with temperature by resp. an increase of the I_{sc} and a decrease of the V_{oc}. Usually the power produced by a solar cell is given by the power produced in its maximum power

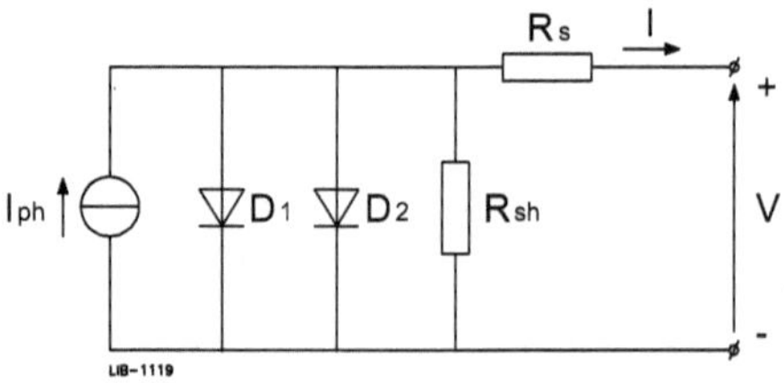

FIGURE 3.2 Equivalent circuit of a solar cell represented in the two-diode model.

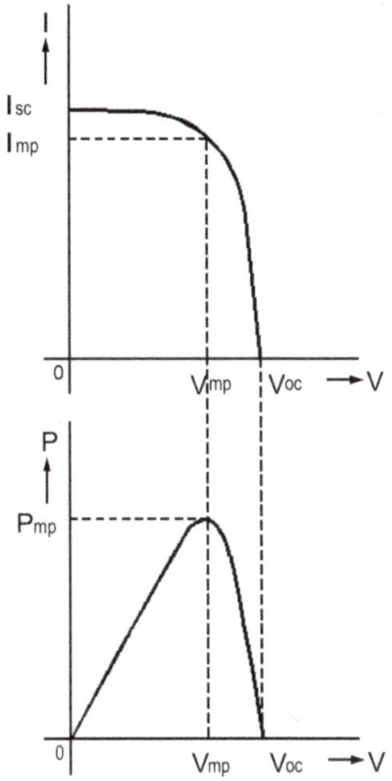

FIGURE 3.3 Above: the current-voltage characteristic also called *I–V* curve of a solar cell. Below: the power-voltage characteristic of a solar cell.

point, P_{mp}, which is the product of the maximum power current, I_{mp}, and maximum power voltage V_{mp}. Also, I_{mp} and V_{mp} are subjected to changes with irradiance and temperature in an almost identical manner as I_{sc} and V_{oc}.

3.3 DESIGN FEATURES OF SOLAR CELLS

Different PV technologies have different design features. In this section, we focus on these design features of the following six commercially available PV technologies: crystalline (m-Si) and polycrystalline solar cells (p-Si); amorphous silicon solar cells (a-Si); DSSCs, that is, the Grätzel Cell; cadmium telluride, CdTe; and CIGS and luminescent solar concentrator PV (LSC PV, which is extensively discussed in Design Case 2 of this book). The opportunities to create different types of designs with these PV technologies can be placed in the context of applications in products, buildings, and vehicles. We explore these opportunities in six different directions: (1) coloring, (2) transparency, (3) flexibility, (4) sizing, (5) two-dimensional design, and (6) three-dimensional design (see also Figure 3.4). This is relevant for each of the three application categories; however, it is most important for solar PV systems in building-integrated PV (BIPV), because of their potential as an aesthetic building material and "shaper" in modern architecture or as ornaments.

In our scope, two-dimensional design refers to: (a) patterning; and (b) shaping of edges, for instance, to create an ellipsoid PV roof. For three-dimensional design, we distinguish between the following two aspects: (a) curvature of surfaces, such

PV technologies	Typical size Substrate (mm x mm)	Patterning	Shaping of edges	Bending and curvature	Color	Transparency
CdTe	Customizable from 10 x 10 to 1,000 x 2,000, rigid and flexible	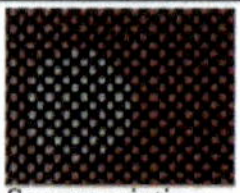 Screen printing front glass	n.a.	 Flexible substrate	Cells are brownish or black. Color by colored front glass	Semi-transparency by wider space between cells and laser scribing
CIGS	Customizable from 10 x 10 to 1,000 x 2,000, rigid and flexible	 Screen printing front glass	n.a.	 Flexible substrate	 Colored front glass	Semi-transparency by wider space between cells and laser scribing
a-SI	Customizable from 10 x 10 to 1,000 x 2,000, rigid and flexible	Screen printing front glass	n.a.	 Flexible substrate	 Cells are usually brownish or black, other colors can be produced	Semi-transparency by wider space between cells and laser scribing
 x-Si/p-Si	156 x 156 (cells), rigid only due to the fragility of the cells	 Cells can be used as pixels	 Laser cutting	 Curvature by rigid carrier substrate	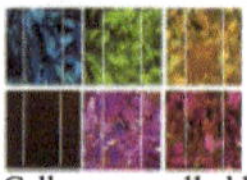 Cells are usually blue, other colors can be produced	Semi-transparency by wider space between cells and punching holes in cells
DSSC	Customizable from 10 x 10 to 1,000 x 2,000, rigid and flexible	Cells can be used as pixels	Defined by cell geometry, i.e., shaping of cell	Flexible substrate	Orange, reddish, purple, depending on the dye applied	Always transparent

FIGURE 3.4 Design features of commercially available PV technologies: CdTe, CIGS, a-Si, m-Si, p-Si, and DSSC. (Adapted from Geenhuizen et al. 2014, and updated with present knowledge.)

as the PV arrays in double-curved car roofs; and (b) spatially distributed structures such as the 500 butterfly-shaped silicon solar cells that occur in the artistic chandelier called Virtue of Blue (Design Case 3 in this book).

For an optimal use of PV technology integrated in products, all elements previously mentioned, which are also shown in Figure 3.4, should be optimally customizable, that is, coloring and transparency, flexibility, sizing, patterning, shaping of edges, bending, and curvature. By changing these features, different types of PV products and PV modules can be created with various appearances and various applications, for instance, tent structures, see-through windshields, roofing elements, facades, window-integrated elements, cars' body parts, and skylights.

3.3.1 Color

Typically, thin-film PV technologies such as CdTe, CIGS, and a-Si have a brownish to black color, which can only be changed by a colored front glass or colored encapsulant. Depending on the dye applied, DSSCs have an orange, reddish, purplish, or greenish color. c-Si solar cells can get a different color by various approaches:

(1) changing the thickness of their antireflection coating, (2) colored front glass, (3) spectrally sensitive front glass, and (4) adding specialty coatings to the cells. These diverse methods can yield all possible colors, however, with a slight loss of efficiency. Finally, the relatively new luminescent solar concentrator PV technology has excellent coloring features due to the presence of specific dyes.

3.3.2 Transparency

All thin-film PV technologies previously mentioned (CdTe, CIGS, and a-Si) can be produced in a semitransparent format by choosing transparent materials for the front and back sheet and by creating wider laser scribes between cells. DSSCs and LSC PV cells are always transparent. Semitransparent silicon PV panels can be created by wider spacing between cells or by punching holes in the cells.

3.3.3 Flexibility

Both thin-film PV technologies and DSSC cells can be produced on flexible substrates (usually made of polymers and sometimes steel), resulting in flexible PV modules that can be bent and which are rollable. Crystalline silicon solar cells are stiff and fragile at the same time and, therefore, can't be produced in a flexible form.

3.3.4 Two-Dimensional Shaping

Silicon solar cells can obtain any shape by laser cutting. Furthermore, graphical patterns can be created on PV modules by screen printing on the front glass. DSSCs can be made in any shape, if they are produced on a flexible polymer substrate. LSC PV cells and mini-modules can be produced in any shape if the LSC consists of a plastic. CIGS, CdTe, and a-Si unfortunately lack proper features to create curved shapes; however, they can be easily produced in any rectangular shape (from very small to very large).

3.3.5 Three-Dimensional Shaping

Silicon PV cells can be bent in PV modules in a limited extent, if they are attached on a rigid substrate that is bendable. This substrate can be glass, metal, or a rigid plastic. Naturally, thin-film PV technologies that have been produced on flexible substrates, as well as DSSCs, have excellent bending features. Here as well, LSC PV cells and mini-modules can be produced in any curved or textured shape if the LSC consists of a plastic.

In any product that contains 3D structures with integrated PV cells, such as shading elements in facades or a complex artistic object like Virtue of Blue (Design Case 3 in this book), self-shading by adjacent cells can occur and should be evaluated in advance during the design process to avoid performance loss under use conditions.

FIGURE 3.5 Possible colors that can be realized with different dyes in luminescent solar concentrator PV applications (From FhG-ISE, 2019.)

3.3.6 Design Features of PV Technologies Summarized

After a comparison of six design features of six different PV technologies, we draw the conclusion that out of the six technologies discussed, silicon-based PV cells, DSSCs, and LSC PV cells provide the best opportunities for product integration (Figure 3.5).

3.4 SOLAR CELLS IN INDOOR ENVIRONMENTS

Some products that contain PV cells have been designed for indoor use. Because indoor irradiance has a different composition than outdoor irradiance (sunlight), it is relevant to be aware about the possible effects of indoor irradiance on the performance of solar cells. Indoor irradiance usually consists of a mixture of natural light (sunlight) that enters a building through windows and other openings, as well as artificial light. Artificial light usually originates from different light sources: lamps that are used to illuminate a house, such as incandescent lamps, fluorescent lamps, and light-emitting diodes (LEDs). In general, indoor lighting conditions are typically created by artificial light sources with irradiance levels below 10 W/m^2 (Müller et al. 2009a, 2009b; Kan 2006; Apostolou 2016). Under these levels of irradiance, solar cells perform at a significantly lower efficiency than the STC efficiency, for which reason they can generate only limited amounts of electricity, in the order of μW up to a few mW/m^2 (Apostolou 2016; Apostolou et al. 2012, 2016).

Several studies have been devoted to solar cell performance under weak light or indoor irradiance conditions. Research on indoor irradiance shows that though indoor irradiance can exceed 500 W/m^2, basic orders of magnitude are typically about 1 to 10 W/m^2 with worst-case scenarios in the winter without the use of artificial light in the range of 0.1 W/m^2 (Müller et al. 2009a, 2009b; Müller 2010).

Many studies on solar cell performance in indoor conditions were also conducted by several researchers in the field (Randall 2003, 2006; Reich et al. 2005, 2009). It was found that under low irradiance conditions of 10 W/m^2, efficiencies of silicon solar cells are in the range of 5% to 10%, depending of their manufacturer or supplier. A recent study deals with the simulation of the performance of various PV materials for different spectral distributions, in order to identify maximum efficiencies under indoor conditions (Freunek et al. 2013). Results of this study show a maximum efficiency of 16% for a GaInP cell under indoor irradiance.

Levels of indoor irradiance were extensively addressed by Apostolou (2016) and Apostolou et al. (2012, 2016), aiming to investigate the relation between indoor irradiance levels and distance of an object from a lighting source—either natural, artificial, or a combination of these two. For that reason, multiple measurements were conducted at various locations in The Netherlands, at rooms with different orientations (South, North), different artificial lighting sources (CFL, LED), and different test setups (Apostolou 2016). It was found that indoor irradiance differentiates broadly depending on the orientation of the room, as well as the type of light sources and the distance from them. Apostolou (2016) shows that indoor irradiance in a south-facing office in The Netherlands during summer ranged between 25 W/m^2 at a distance of 0.5 m from the window and around 5 W/m^2 at a distance of 4 m from the window (see Figure 3.6). In a north-facing office, indoor irradiance ranged between 17 W/m^2 at 0.5 m from the window and 3 W/m^2 at 4 m, respectively (see Figure 3.7). The use of artificial lighting in the rooms appeared to be important, especially at a distance of more than 2 m from the window, where natural light indoors is not sufficient. Artificial lighting helps to keep indoor irradiance more stable away from the windows. However, indoor irradiance based on artificial lights usually ranges between 3 and 7 W/m^2, which is only sufficient for low-power PV products.

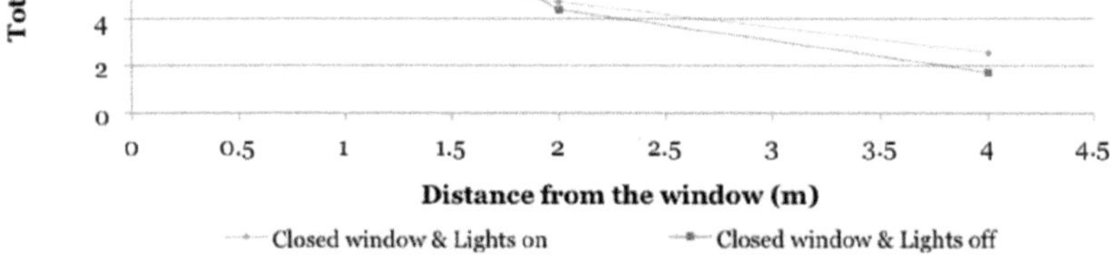

FIGURE 3.6 Total measured irradiance (W/m^2) at a south-facing office on June 10, 2015, at Delft, The Netherlands. In both cases, the window was closed during the measurements, while the lights were on for the blue line and off for the red one.

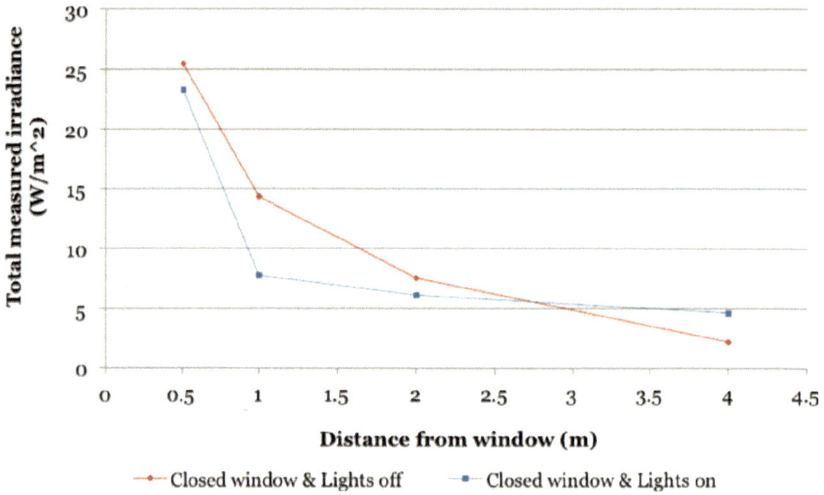

FIGURE 3.7 Total measured irradiance (W/m^2) at a north-facing office on June 10, 2015, at Delft, The Netherlands. In both cases, the window was closed during the measurements, while the lights were on for the blue line and off for the red one.

The values of indoor irradiance previously mentioned cannot be considered fixed because they are strongly influenced by the latitude and longitude of the room, the season (e.g., winter, summer), the weather conditions (e.g., sunny, cloudy, rainy), the use of artificial lighting (amount of lamps, type of lights), objects and shad(ow)ing at the indoor environment, distance from windows and artificial light sources, type of glazing, etc. Therefore generally, it is considered that indoor irradiance typically ranges between 1 and 10 W/m^2. However, this does not exclude the possibility of having higher indoor irradiance during a summer day and quite lower irradiance on a winter day. Generally, indoor irradiance is higher near the window and lower while the distance from the window increases.

REFERENCES

Apostolou G. (2016). Design features of product-integrated PV: An evaluation of various factors under indoor irradiance conditions, PhD Thesis, Industrial Design Engineering Department, Technical University of Delft (TU Delft), Delft, the Netherlands.

Apostolou G., Reinders A.H.M.E., Verwaal M. (2012). Spectral irradiance measurements in a room fit for indoor PV products, *Proceedings of the 27th European PV Solar Energy Conference (EUPVSEC)*, Poster Presentation, Frankfurt, Germany, pp. 4240–4244.

Apostolou G., Reinders A.H.M.E., Verwaal M. (2016). Comparison of the indoor performance of 12 commercial PV products by a simple model, *Energy Science & Engineering*, Wiley Online Library, January 2016, Article first published online: January 22, 2016, Volume 4, Issue 1, pp. 69–85.

FhG-ISE (2019). Fraunhofer Institute of Solar Energy Systems, Photovoltaics Report, www.ise.fraunhofer.de, visited June 2019.

Freunek (Müller) M., Freunek M., Reindl L.M. (2013). Maximum efficiencies of indoor photovoltaic devices, *IEEE Journal of Photovoltaics*, 3(1), 59–64.

Geenhuizen M. van, Schoonman J., Reinders A.H.M.E. (2014). Diffusion of solar energy use in the built environment supported by new design, *Journal of Civil Engineering and Architecture*, 8(2): 253–260.

Green M.A. (1982). *Solar Cells: Operating Principles, Technology and Systems Applications*, Prentice Hall, Englewood Cliffs, NJ.

Green M.A. (2003). *Third Generation Photovoltaics: Advanced Solar Energy Conversions*, Springer, Berlin, Germany.

Kan S.Y. (2006). SYN-Energy in solar cell use for consumer products and indoor applications, PhD dissertation, Final Report 014–28–213, NWO/NOVEM, Technical University of Delft, Delft, the Netherlands.

Müller M. (2010). Regenerative Mikroenergiesysteme am Biespiel von Photovoltaik in Gebauden, PhD Thesis, Der Andere Verlag, Osnabruck, Germany.

Müller M., Hildebrandt H., Walker W.D, Reindl L.M. (2009b). Simulations and measurements for indoor photovoltaic devices, In: *Proceedings of the 24th European Photovoltaic Solar Energy Conference*, Hamburg, Germany, pp. 4363–4367.

Müller M., Wienold J., Walker W.D., Reindl L.M. (2009a). Characterization of indoor photovoltaic devices and light, *Proceedings of the 34th IEEE Photovoltaic Specialist Conference*, Philadelphia, PA, pp. 738–743.

NREL (2019). National Renewable Energy Laboratories, Best Research-Cell Efficiency Chart. https://www.nrel.gov/pv/cell-efficiency.html, Last visited July 2, 2019.

Randall J.F. (2003). On the Use of Photovoltaic Ambient Energy Sources for Powering Indoor Electronic Devices. PhD Thesis, Lausanne, Switzerland.

Randall J.F. (2006). Designing Indoor Solar Products: Photovoltaic Technologies for AES, PhD dissertation, ETHZ Switzerland, John Wiley & Sons, London, UK.

Randall J.F., Jacot J. (2003). Is AM 1.5 applicable in practice? Modeling eight photovoltaic materials with respect to light intensity and two spectra, *Renewable Energy*, 28, 1851–1864.

Reich N.H., van Sark W.G.J.H.M., Alsema E.A., Kan S.Y., Silvester S., van der Heide A.S.H., Lof R.W., Schropp R.E.I. (2005). Weak light performance and spectral response of different solar cell types, *Proceedings of 20th EUPVSEC*, Barcelona, Spain, pp. 2120–2123.

Reich N.H., van Sark W.G.J.H.M., Alsema E., Lof R., Schropp R., Sinke W., Turkenburg W. (2009). Crystalline silicon cell performance at low light intensities, *Solar Energy Materials & Solar Cells*, 93, 1471–1481.

Reinders, A.H.M.E., Verlinden, P.J., Van Sark, W.G.J.H.M., Freundlich, A. (2017). *Photovoltaic Solar Energy From Fundamentals to Applications*, Hard-cover book, John Wiley & Sons, London, UK.

Design Case 3

Virtue of Blue

Jeroen Verhoeven and Angèle Reinders

3.1 INTRODUCTION

In 2009, renowned Dutch design studio DeMakersVan designed and realized a beautiful, large solar-powered chandelier, measuring over 1440 × 1440 × 1620 mm, which was presented to the public during a showing of their collection in the London-based Blain Southern Gallery (https://www.blainsouthern.com/exhibitions/virtue-of-blue) in the second quarter of 2011. Thereafter, the chandelier has been exhibited in many places among which are Museum Boijmans van Beuningen in Rotterdam and Droog in Amsterdam. At present, it is in the permanent collection of the Corning Museum of Glass (https://www.cmog.org/artwork/virtue-blue) in New York, USA. In this design case, the vision behind this design, its practical realization, and estimates of the power production in relation to available irradiance will be presented (Figure DC-3.1).

(a) (b)

FIGURE DC-3.1 (a) Virtue of Blue in an indoor environment, and (b) details of this object.

FIGURE DC-3.2 Initial visual impression of Virtue of Blue in 2009.

3.2 VISION

Self-powered by solar energy, Virtue of Blue (https://www.aleinno.com) is a chandelier that playfully explores an economy of light through innovative materials, such as silicon solar cells, hand-blown glass, aluminum, and stainless steel.

The piece is intrinsically self-sustaining as it absorbs the energy of daylight to fuel its own illumination. More than 500 cells have been cut into the shapes of three different breeds of butterflies, and these seem to flutter around a central flame-like hand-blown glass bulb, their iridescent wings glinting in the light. The semiotics of this design are highly significant as the butterflies become signifiers of the light's self-sufficiency (Figure DC-3.2).

Physically, these insects also power their own bodies, using their wings to absorb the sunrays, in turn raising and sustaining their own body temperatures to that which is necessary for their survival.

3.3 PRACTICAL REALIZATION

The practical realization of this chandelier that acts as a self-sustaining lighting device was a challenge. Despite the highly complex design of this lamp, it appeared possible to realize it, however only by several smart design solutions and numerous ray-tracing simulations of indoor irradiance and solar-power conversions by overlapping solar cells. The resulting design was prototyped by laser cutting and by customized electrical connections between solar cells (Figure DC-3.3).

FIGURE DC-3.3 Final design of Virtue of Blue as realized in 2011.

3.3.1 List of Demands

In the first stage of the design process, the functions that have to be fulfilled by Virtue of Blue had to be identified. These functions can be translated in a list of demands valid for the whole chandelier, which are as follows:

- The solar chandelier has to emit enough light to be visible as a decorative light for five hours a day in a room with other lights.
- The solar chandelier has to be interpreted as an autonomous object that generates its own electricity.
- The solar chandelier needs to have dimensions that suit the space of an average museum hall.
- The butterflies have to be very thin; they have to approach the delicateness of a real butterfly as much as possible.
- The multicrystalline solar cells are a very important for the aesthetics of the chandelier. The multicrystalline structure with its distinctive reflections has to be visible, and the blue color of the cells has to remain unaltered.

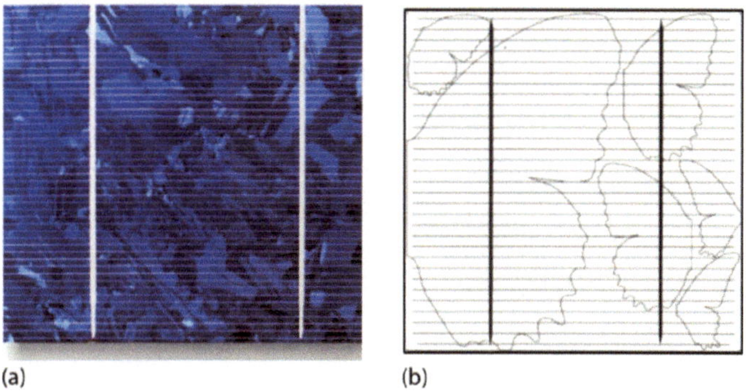

FIGURE DC-3.4 (a) Multicrystalline solar cell used in Virtue of Blue, and (b) projection of the wings of the butterflies on a solar cell ready to be cut out.

FIGURE DC-3.5 Different shapes of butterflies applied in Virtue of Blue in practice in three different sizes.

3.3.2 Mechanical Design

The solar cells used in Virtue of Blue are multicrystalline cells produced by Sunways (Figure DC-3.4). These are cut in the shape of butterfly wings. There exist three basic shapes (see Figure DC-3.5), and each shape is applied in four different sizes: A, B, C, and D (with A the largest and D the smallest). For the attachment of the cells to a frame, a sandwich structure was developed, which contained a top cell attached to an aluminum carrier and a non-functional bottom cell also attached to this carrier. Both sides of the resulting silicon butterfly wings have been coated to protect them against corrosion (Figure DC-3.6). The final design of Virtue of Blue consists of three helix structures that hold the swarms of butterflies (Figure DC-3.7).

3.3.3 Further Components and Functioning

The lighting function of Virtue of Blue is provided by LED strips that are powered by batteries, which are charged during the day by the power generated by

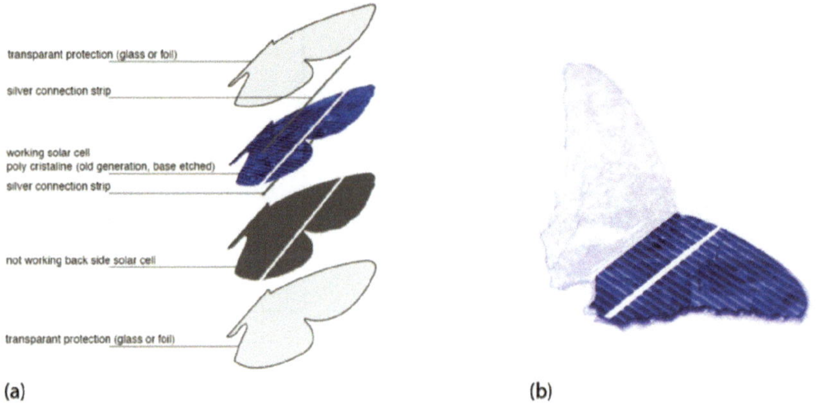

FIGURE DC-3.6 (a) Sandwich structure for PV butterflies, and (b) prototype showing a butterfly solar cell covered and the bare aluminum carrier.

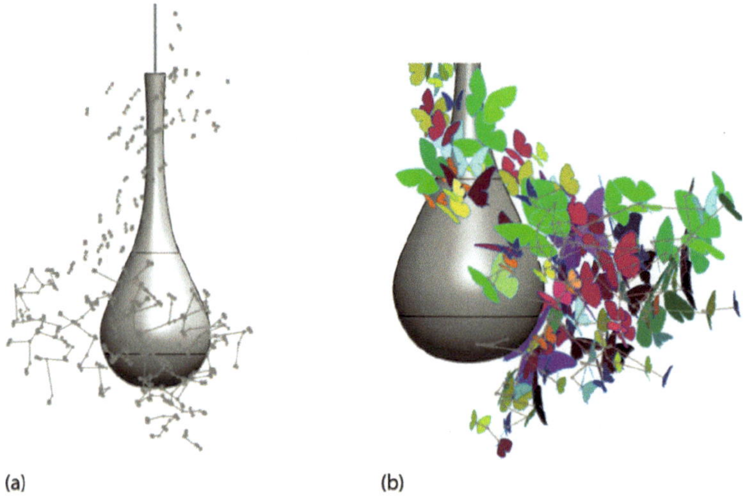

FIGURE DC-3.7 (a) Layout of one of the bare frames, which can hold and connect the solar butterflies; and (b) enlargement of the frame with butterflies attached to it.

the solar cells in the butterfly wings. These components are located in the interior of the glass bulb in the middle of the chandelier. When the solar chandelier is placed indoors, on some days the power produced by the solar cells won't be high enough to charge batteries to their full capacity; therefore, the decision was made to use an additional external power source for the solar chandelier.

3.4 IRRADIANCE ESTIMATES IN RELATION TO THE DESIGN OF VIRTUE OF BLUE

For the purpose of irradiance estimates in various surroundings (outdoors and indoors), simulations were executed using the software packages Dialux and 3ds Max.

To determine the energy yield of Virtue of Blue, images are rendered containing the hourly received amount of illuminance on each wing of the solar cell butterflies. From this information, using a conversion factor that is location specific, the average hourly irradiance could be calculated. For a worst-case scenario in London, on the shortest day of the year, namely the 21st of December, the energy yield of Virtue of Blue was 5.8 Wh per day while positioned behind a window in an indoor space. This implies that on other days the energy yield would be much higher.

Based on the simulation results, the interconnection scheme between solar butterflies of various size was optimized while maintaining the full aesthetics of the chandelier.

Virtue of Blue got divided in three nearly identical sections, out of which two are shown in Figure DC-3.8. Each section contains a circuit of 29 butterflies at a maximum. This takes advantage of inherent symmetries present in the design of Virtue of Blue and enables the solar chandelier to profit from irradiance from different sides.

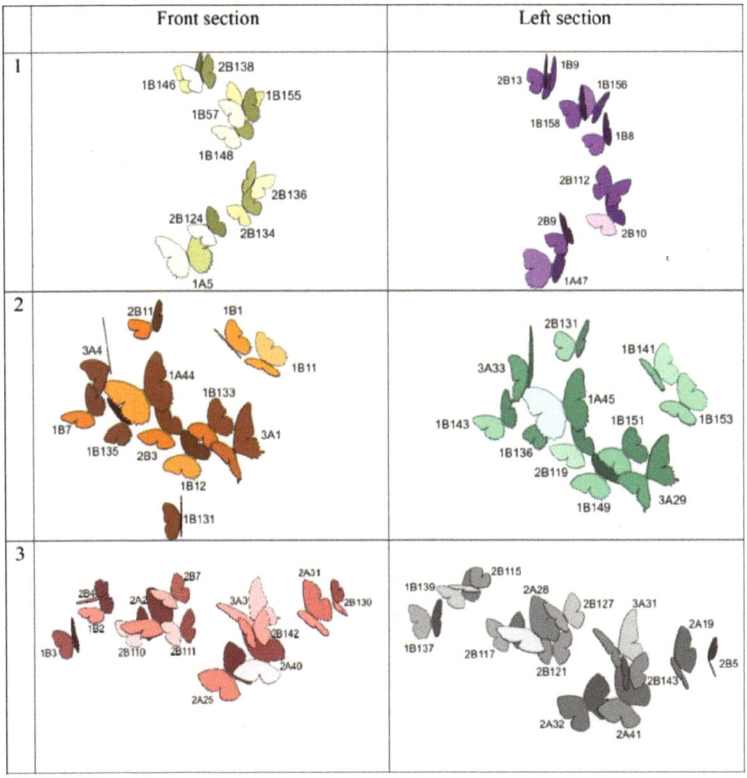

FIGURE DC-3.8 Two of the sections which contain functional solar butterflies in Virtue of Blue.

ACKNOWLEDGMENTS

We would like to acknowledge those who contributed to the realization of Virtue of Blue, in particular, Kay van Mourik, Anniek Braham, Sebastian Kettler, and Rik de Konink

4 Product-Integrated PV

Georgia Apostolou and Angèle Reinders

CONTENTS

4.1 INTRODUCTION

Each year new solar cell technologies are spreading the markets. Solar cells with higher efficiencies, better characteristics, are ready to be used in various products and applications. The most well-known applications of photovoltaic (PV) technologies are PV systems (either grid-connected or stand-alone) and so-called building-integrated PV (BIPV). BIPV will be further discussed in Chapter 5 of this book. However, since the 2000s, PV cells became more frequently applied in products, in so-called product-integrated PV (PIPV). Nowadays, almost 20 years after the first PV-powered products, a range of new applications for PV cells exists, some quite useful, well-designed and practical for users. Therefore, this chapter explores the field of PIPV, a term used for all types of products that contain solar cells in one or more of their surfaces, aiming at providing power during operation of a product, otherwise said during product use. PIPV began to be widely introduced around 2000, although the use of PV cells in products dates back to the 1960s (see Chapter 1).

PIPV includes products such as PV-powered boats, aircrafts, cars, bicycles, camping tents, street lights, recycling bins, decorative lights, PV-powered watches, calculators, PV-powered lamps, sensors, chargers, toys, low-powered kitchen appliances, entertainment appliances, furniture, or PV-powered art objects. The incorporation of PV systems in products could offer various benefits, such as enhanced functionality of the product as a result of energy autonomy, and independence and freedom of use due to the absence of a connection to the electricity grid, as well as the opportunity to reduce the capacity of batteries in portable products and therefore making them more sustainable. Furthermore, PV products represent a very reliable solution for the supply of electricity in areas that lack access to an electricity grid.

The first three chapters of this book offered some general information regarding the origin of PV-powered products and their development in the course of time, as well as the PV technologies and their design features. This chapter will use the knowledge introduced in the previous chapters to present, describe, and analyze the various categories of PIPV. In the next sections of this chapter, existing PIPV will be presented and classified in categories based on their use, power, and technical features. As well as the environmental aspects, human factors and cost of PIPV will be addressed. More specifically, this chapter presents an overview of the design features and characteristics of 105 PV-powered products that were on the market from 2011 to 2019, covering an analysis of concerning of their power range, PV technologies used, battery technologies, manufacturers, place of use, sales price, and several other design features[1] (Apostolou 2016).

First though, the criteria for the characterization of a product such as a PIPV product (Reinders and van Sark 2012; Reinders and Apostolou 2017) are the following:

1. The existence of an integrated PV cell on a product's casing or another surface
2. The use of the energy that is generated by the PV cells for the energy requirements of the product, during its operation
3. The option of straightforward user interaction with the product
4. Energy storage in batteries or another storage device
5. Use of the product on land
6. Easy transportation of the product

PV systems and PIPV have many differences related to the range of the energy produced, as well as their size, design, technical features, cost, way of manufacture, life cycle, and other aspects (see Table 4.1). However, the main difference between them is that PIPV also includes product parts such as plastics, iron, or glass parts, which consist of the housing (outer casing) of the products (Reinders and van Sark 2012).

[1] Apostolou G. (2016) presents in her PhD Thesis entitled "Design Features of Product-Integrated PV: An Evaluation of Various Factors under Indoor Irradiance Conditions," an analysis of 90 PV-powered products. This analysis was carried out from 2011 to 2013 at Delft University of Technology in The Netherlands. An updated version of this initial selection of 90 PV products is presented here, enriched by 15 more PIPVs, launched in the markets from 2016 to 2019.

TABLE 4.1
Differences between PV Systems and PIPV Summarized

	PV Systems	PIPV
Energy produced	From mW up to GW	From μW up to a few kW
Size	Depends on the available area (m^2 to km^2)	Depends on the available area (cm^2 to m^2)
Design	Simple	More complex
Technical features	PV modules and inverters, monitoring systems	PV cells, batteries, dedicated customized electronics
Life cycle	Around 20 years	Around 5 years

Source: Apostolou, G., Design Features of Product-Integrated PV: An Evaluation of Various Factors under Indoor Irradiance Conditions, PhD Thesis, Industrial Design Engineering Department, Technical University of Delft (TU Delft), Delft, the Netherlands, 2016; Apostolou, G. and A. H. M. E. Reinders, *Energy Technol.*, 2014.

Furthermore, PIPV includes additional PV system components, as well as mechanical and electrical parts (PV cells, batteries, inverters, controllers, etc.). On the one hand, PV systems produce electricity for a wide range of purposes from microwatts to gigawatts, while on the other hand PIPV is used in a product context (Reinders and van Sark 2012; Kan et al. 2006; Randall 2005; Reinders 2002; Reinders et al. 2012; Timmerman 2008; Veefkind et al. 2006). PIPV provides multiple functions for products that require powering of electronic appliances, lighting, sound, telecommunications, heating, cooling, or transportation. Moreover, with PIPV users are able to interact directly with the products. The functionality of the product depends on user habits for charging, as well as the usage profile of the product (Reich et al. 2008).

Another difference between PV systems and PIPV is their lifetime. PIPV usually have a shorter life than PV systems, which is limited to a few years (up to five years), because consumer products are intended to be used for a short period of time. In contrast, PV systems are designed to reliably function for longer periods of time, typically for more than 20 years, with the purpose of continuously providing electricity. As a consequence, PIPV is considered as a distinct category in the wide variety of PV applications, and therefore it is studied separately.

While designing a PV-integrated product (Randall 2005; Reinders et al. 2012; Timmerman 2008; Veefkind et al. 2006; Alsema et al. 2005; Reich et al. 2009), it is vital not only to use an alternative source of energy, but also to use inspiration from everyday products, as well as to be aware of the daily energy consumption of a product in its context of use. The product should suit the user's lifestyle, while meeting his energy requirements at the same time. In this regard, during the analysis process not only the technical function of a product should be involved, but also the psychological, social, economic, and cultural functions that a product should accomplish, as well as users' sustainable behavior (Bakker et al. 2010; Jong and Maze 2010; Keyson and Jin 2009; Wever et al. 2008). Customer satisfaction and user interaction with a product appeared to have a high significance and should be also included in the design process (Jong and Maze 2010; Scott 2008, 2009). Consumers' satisfaction is an indication of by what means products and services delivered by a company approaches or surmounts users' expectations. It is a quite ambiguous and abstract concept and varies from person to person and product-service combination to product-service combination.

Users' benefits, derived from the incorporation of PV systems in products (Reinders and van Sark 2012) can be summarized as the enhanced functionality of the product offering energy stability, security, independence, freedom due to the absence of the connection to the electricity grid, and extra autonomy in batteries, which can be achieved by the reduction of their extended use.

In an attempt to introduce functionality and ecological behavior to our lives, we target energy savings in various ways. This approach begins from simple things in everyday life that eventually are those that make the greatest difference over time. Therefore, product developers search for natural solutions; ecological products (eco gadgets), which are practical and have a multipurpose design for use inside and outside the house. This may constitute an important application of energy independence, whether this applies to a calculator or a car.

Because of the rich diversity of PIPV, its categorization is further explained and detailed in Section 4.2 presenting an overview of existing product categories in PIPV. Subsequently, in Section 4.3, the design features of 105 PIPV (Apostolou 2016) are discussed. In Section 4.4, the environmental aspects of PV technology in products are addressed, and in Section 4.5, human factors through users' experiences and profiles. In Section 4.6, costs of PIPV are shortly discussed. The chapter finishes with conclusions in Section 4.7.

4.2 OVERVIEW OF EXISTING PIPV

As mentioned in Section 4.1, nowadays PV solar cells are utilized in various different product categories. At the moment, the following product categories of PIPV are recognized (Reinders and van Sark 2012; Reinders and Apostolou 2017):

1. Consumer products (see Figure 4.1a–d, g)
2. Lighting products (see Figure 4.1e–f, h–i)
3. Business-to-business applications (see Figure 4.1j–l)
4. Recreational products (see Figure 4.1p)
5. Vehicles and other forms of transportation (see Figure 4.1m–o)
6. Arts (see Figure 4.1q)

The main features of these categories and examples of PV products in them are addressed in the next paragraphs of this section.

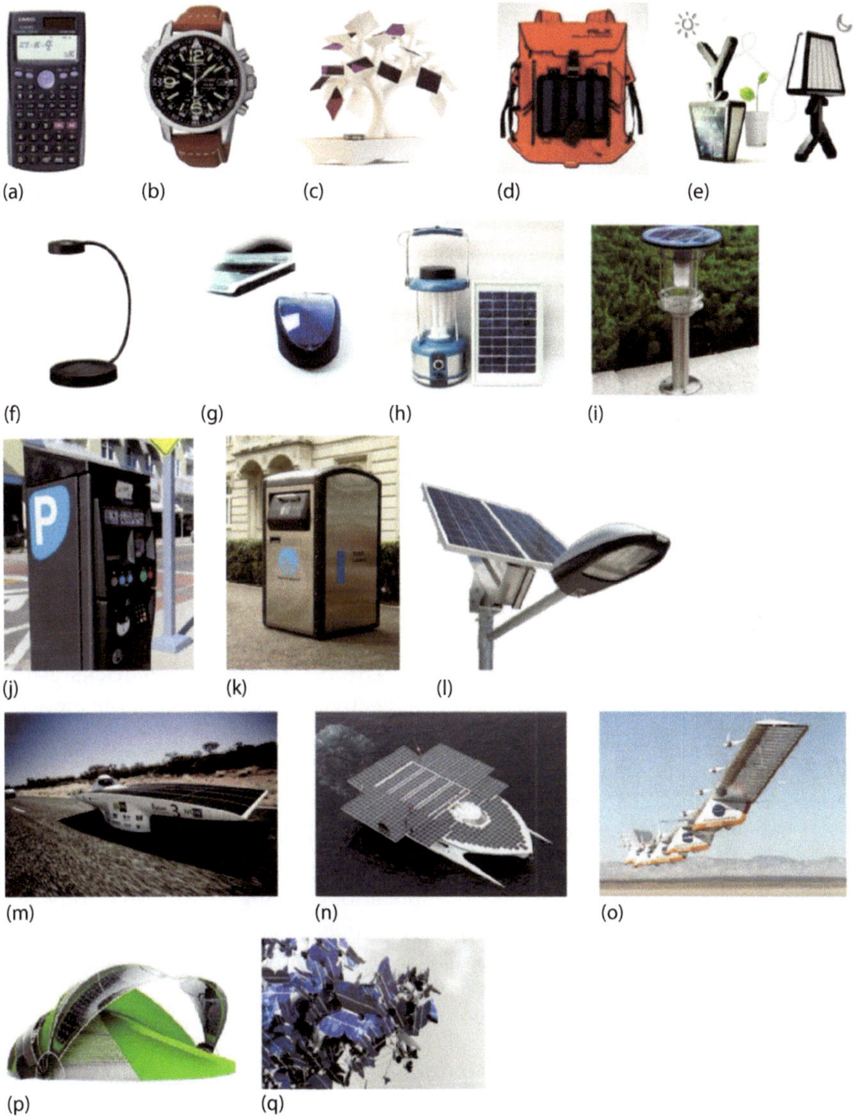

FIGURE 4.1 Photovoltaic products of various product categories: (a) solar calculator, (b) solar watch, (c) iPhone charger by Vivien Muller, (d) solar-powered bag, (e) Spark lamp, (f) Ikea Sunnan lamp, (g) SoleMio (Photo by Matthijs Netten), (h) solar lantern, (i) solar garden light, (j) solar-powered parking meter in Virginia, (k) automated trash bin Big Belly, (l) solar traffic light, (m) solar-powered car "Nuna" (n) Planet Solar Catamaran, (o) Helios solar aircraft, (p) solar-powered tent, and (q) PV-powered chandelier. (From Apostolou, G. and Reinders, A.H.M.E., *Energy Technol.*, 2014.)

4.2.1 Consumer Products with Integrated PV

To the category "consumer products" belong products of everyday use with a PV cell's power ranging from 0.1 mW up to 10 W (Reinders and van Sark 2012; Apostolou 2016). PV cells are applied to products such as toys with integrated PV cells, kitchen appliances, mobile phones, PV chargers used in cell phones (see Figure 4.2), lighting devices—which constitutes a large category by itself—portable consumer electronics, PV-powered thermometers, radios, calculators, PV-powered watches, MP3 players, PV headsets, automated lawn mowers, solar sensors (see Figure 4.3a and b), and Internet of Things (IoT) applications (see Figure 4.3c and d).

Sensors and IoT applications are recently used in a more extended scale than in the past. They can be used separately or even combined. Sensors are devices, which can detect events, motions, or changes in their environment, for example, a temperature or humidity difference, change in levels of irradiance, smoke detection, or motion. The events/information they record can be sent to other electronics, such as a computer processor or a smartphone. Sensors are broadly used in outdoor lighting systems, such as garden lights, in security cameras, or in indoor lighting systems, thermometers, temperature indicators, smoke or fire detectors, and many other home applications.

IoT applications are a new product category that was officially launched between 2008–2009. IoT describes devices that we use in our daily life, which have the ability to be Internet connected (Brown 2016a, 2016b; ITU 2015; Hendricks 2015). IoT applications/devices include Internet connectivity, hardware (e.g., sensors) and electronics. They can communicate and interact with other applications/devices and/or users via Internet and be remotely monitored and/or controlled (Brown 2016a, 2016b; ITU 2015; Hendricks 2015). IoT technology is broadly used in "smart homes" and buildings, with applications such as lightings, cameras, heating, air-conditioning, security systems, and media that can be monitored and controlled remotely, through smartphone applications (Kang et al. 2017; Business Insider 2017). IoT consists of a new market, which is quite promising during the next years. A few IoT applications and solar sensors are included in the analyzed

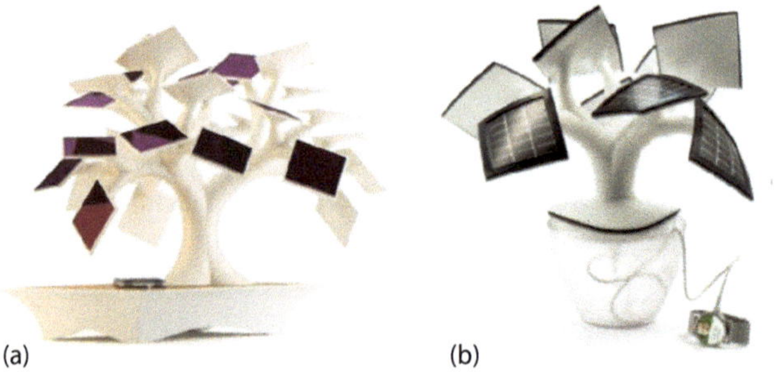

(a) (b)

FIGURE 4.2 iPhone chargers (a) Electree, (b) and Electree mini by Vivien Mueller. ©Vivien Mueller.

FIGURE 4.3 (a) Solar-powered sensor LED wall light, (b) solar-panel-powered LoRaWAN RTU for SDI-12 sensors-TBS12S by copyright TekBox, (c) energy-harvesting PMIC targets solar-powered wireless sensor, and (d) Bluetooth solar indoor development kit (LES100) copyright PowerFilm Solar.

PV products that are presented in this chapter. Figure 4.3 illustrates a couple of solar sensors and IoT applications, which are included in the list of the analyzed PV products.

A well-designed and styled PV product was described in Design Case 1, after Chapter 1. An interesting product idea for a PV-powered consumer product is the computer mouse Sole Mio (see Figures 4.1g and 4.4), which had been analyzed, prototyped, and tested during the Dutch SYN-Energy project (Reich et al. 2007, 2008a, 2008b, 2009; Alsema et al. 2005). Unfortunately, this product is not yet commercially available. Figure 4.5 presents several typical categories of PV products. All consumer PV-powered products, including PV-powered lighting, form a rather large share of all PIPV analyzed in this chapter (around 67%).[2]

[2] All graphs of Chapter 3 are adapted by Apostolou and Reinders (2014) and updated by Apostolou in 2019.

FIGURE 4.4 The PC computer mouse Sole Mio, analyzed, prototyped, and tested during the Dutch SYN-Energy project. (From Reich, N.H. et al., A solar powered wireless computer mouse: Design, assembly and preliminary testing of 15 prototypes, *Proceedings of 22nd European Photovoltaic Solar Energy Conference*, Milan, Italy, 2007; Reich, N.H. et al., *Solar Energy*, 83, 202–210, 2009; 2008a, 2008b, Photo credits Matthijs Netten.)

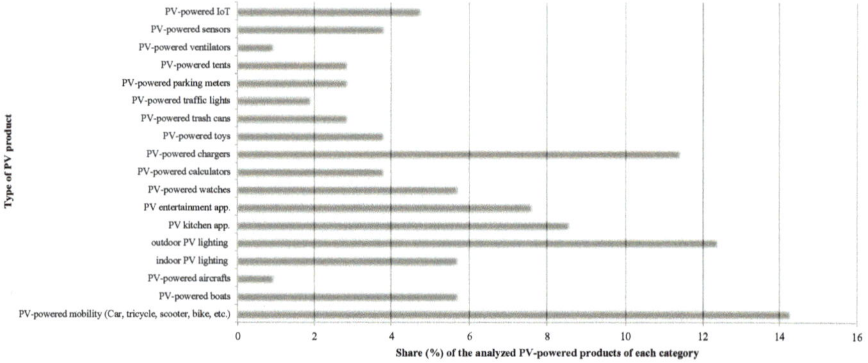

FIGURE 4.5 Share (%) of PV products for each product category in our evaluation of PIPV.

4.2.2 Lighting Products with Integrated PV

Numerous self-powered lighting products such as flashlights, ambient luminaires (see Figure 4.6), lamps for bicycles, garden lamps, pavement lights, indoor desk lamps, street lighting systems, and other products for lighting of public spaces are commercially available. The power of lighting products varies between 1 W for one

FIGURE 4.6 Orkys by Vivien Mueller. Orkys is a solar-powered orchid light. It captures energy in its "leaves" during the day to light its flowers during the night. ©Vivien Mueller.

light emitting diode (LED) and 100 W, which could be the power of a street light pole (Reinders and van Sark 2012). In Figure 4.5, 19 lighting products with integrated PV both for indoor and outdoor use are exposed; this is 18% of the total share of PIPV evaluated in this study. Thirteen PV lights among 19 were used outdoors, such as garden lights, torches, or streetlights, and 6 PV lamps were used indoors, mainly consisting of desk lamps.

A solar lamp is a portable light device composed of an electric light source—usually an energy-efficient one, such as a fluorescent lamp or an LED—an integrated PV panel, and a rechargeable battery. Outdoor lamps are typically used for garden decorations, while indoor solar lamps are often used for general illumination. Their function is based on the operating principle that they recharge during the day, they light at sunset (automatically or using a switch), and remain powered during the night, depending on the amount of sunlight that they received during the day. Discharging time varies from one type of lamp to another; however, it usually ranges from 8 to 10 hours.

4.2.3 PIPV in Business-to-Business Applications

In the same way as consumer and lighting products, PIPV has also been applied in business-to-business (B2B) applications. The power of solar cells of B2B products varies between 10 W to 200 W (Reinders and van Sark 2012). Examples of business-to-business applications are parking meters, traffic control systems, traffic lights, and trashcans. Nowadays, PV-powered public trashcans, with automated control of trash collection, are available and they are used successfully in many cities (Philadelphia, Newton, Shanghai). Their function centers on crushing trash, using electricity produced by integrated solar panels. This offers a significant increase in bin capacity, by carrying approximately five times more waste or recycling compared to traditional trash bins.

Another B2B application available on the markets is the PV-powered ventilator. PV ventilators can be operated in static boats and parked cars. The PV ventilator helps to decrease the temperature and humidity in the car, by withdrawing the hot stuffy air from inside the car and replacing it with fresher air from outside. PV also used in products for telecommunication (e.g., radiotelephone systems, microwave telephone and television repeaters), security, and environmental monitoring. In Figure 4.5, nine PV products (almost 8.5%) of the category "business-to-business" applications are included in the set of the 105 PV products analyzed in this chapter.

4.2.4 Recreational Products with Integrated PV

Products in the power range of 50 to 500 W (Reinders and van Sark 2012) that belong to this product group are the following: PV-powered caravans, motorhomes, campers, tents, solar-powered pond equipment (e.g., pond lights), solar-powered fountains, and PV products for water sports (e.g., underwater lens). Nowadays, modern caravans, motor homes, or solar tents are gradually beginning to PV panels attached to their roof, which enables the use of lighting, refrigeration, laptop charging, and entertainment equipment, far away from the grid connection and supports battery charging. In Figure 4.5, three recreational products with integrated PV were analyzed, concerning three different types of solar-powered tents.

4.2.5 Arts with Photovoltaics

The category "arts" contains mainly decorative products. The PV power of art products can vary significantly—from several mWs to some kWs—as can the location of usage (indoors and outdoors) (Reinders and van Sark 2012). Some examples of "art" PV products are the following: arty objectives (e.g., PV jewelry), art for public spaces (e.g., statues, fountains, art constructions for decoration of parks or squares), and indoor art (e.g., a PV-powered chandelier, decorative PV lights). In this category, aesthetics are combined with usefulness. Well-formed/built constructions bring imagination to life and present a beautiful visual outcome, due to the special PV features (color, flexibility, reflections). The solar chandelier Virtue of Blue (see Figure 4.7) is a design case in this book.

FIGURE 4.7 The solar-powered chandelier Virtue of Blue. (Courtesy of Jeroen Verhoeven.)

4.2.6 Vehicles and Other Forms of Transportation

This category includes bikes, boats, cars, and planes from 200 W power to 1 kW for cars, 1 to several kWs for boats, and numerous kilowatts for planes (Reinders and van Sark 2012). Twenty-two PV-powered vehicles were analyzed (Apostolou and Reinders 2012) (Figure 4.5): 15 PV-powered means of transport, such as PV-powered cars, electrical tricycles, e-scooters, and e-bikes; 6 PV-powered boats; and 1 PV-powered aircraft. The category of solar cars includes solar racing cars, bicycles, and golf cars. Most of the above PV applications are still in the demonstration phase, and they are not commercially available yet. Main drivers for product development in this category are contests like the Solar Challenge (World Solar Challenge 2020) for racing cars in Australia and for solar-powered boats in The Netherlands, for which reason new PV vehicles are designed, produced, and investigated for potential future market implementation. Ideas and thoughts regarding how to design a solar racing car are given by Solar team UT in a design case in this book.

4.2.7 Categories of Indoor PV Products

Of product categories outlined above, the majority of products are mainly high-power PV products designed for outdoor use. This is due to the amount of energy that they need to function properly, which makes it rather difficult for them to be efficient indoors. The share of outdoor PV products in the sample that is investigated here is 62%, while the share of indoor PV products is 38%. Turning to the indoor environment, where indoor irradiance with typical levels between 0.1 and 10 W/m^2 is significantly lower than outdoors (usually up to 1000 W/m^2), fewer products are available. Therefore, different product categories are modulated for indoor use. The low-power PV product categories for indoor use range between 0.1 mW up to a maximum of 10 W, and they are defined as follows:

1. Consumer products (e.g., PV-powered toys, calculators, watches, entertainment applications, kitchen appliances, PV-powered phone chargers for indoor use, sensors, IoT applications)
2. Lighting products (including low-power desk lamps)
3. Art objectives (objets d'art) (requiring low energy supplies)
4. Furniture

The operational voltage of low-power PV products for indoor use typically ranges from 1 to 5 V. Most indoor PV products (around 61%) are completely autonomous; they do not require batteries for supplementary energy storage, and they are not connected by a socket to the grid. They function exclusively using the electricity produced by the PV cells they wear. A specific case of a PV-powered table is a good example in the category of indoor PV products. It is therefore shown as a design case in this book.

4.3 SYSTEM DESIGN AND ENERGY BALANCE

In this section, a selection of 105 PV products is analyzed regarding their design features. For this purpose, a previously developed list of 90 PV products (Apostolou 2016; Apostolou and Reinders 2014) has been updated and expanded with 15 more products. This was necessary because of limited amount of indoor and low-powered PV products of the first list, as well as the need to add new commercially available PV products from 2014 to 2019. Some of the analyzed products are included in Figure 4.1.

The selected 105 PV products were found on the Internet during a research of commercially available PV products from 2011 to 2019. Thus, 90 out of 105 of the analyzed products have been investigated in the framework of the course "Smart Energy Products" of Sustainable Energy Technology (SET) at Technical University of Delft during the years 2011–2012 and 2012–2013 (Apostolou 2016; Apostolou and Reinders 2014). In addition, 103 master's students in 2011–2012 and 21 students in 2012–2013 participated in the course "Smart Energy Products." Figures of the 90 PV-powered products were distributed to the students together with a questionnaire they were asked to answer.

The questionnaire included questions regarding the PV cell of the product (e.g., technology, area, PV power), the battery (e.g., technology, capacity), the cost, product's functions, dimensions, and other technical data. The information was collected through the network, the products' data sheets if available on the Internet, or by contacting the manufacturer. After receiving the initial information by the students, the author executed a more detailed analysis and research for the missing information.

Technically speaking, four PV system categories can be distinguished for PIPV (Apostolou 2016; Apostolou and Reinders 2014), (see Figures 4.8 and 4.9):

1. Autonomous PV system including battery
2. Chargeable PV system including battery
3. Autonomous PV system excluding battery
4. Autonomous hybrid PV system including battery

A hybrid PV system is a system consisting of two or more renewable energy sources (i.e., solar and wind energy), batteries, and/or a connection to the grid. In the case of an autonomous hybrid PV system including battery, as Category 4 indicates, the system uses PV cells and a small wind generator, which are used together to provide high system efficiency and energy balance. Typically, outdoor lighting poles can have this system configuration. In Figure 4.8, the fourth scheme illustrates the autonomous hybrid PV system category including battery. RET 2 in the scheme indicates the additional renewable energy technology that is used together with the PV cells.

Not surprisingly, about 71 out of 105 PV products analyzed belong to the autonomous PV systems with batteries (see Figure 4.9). It is remarkable that 19 PV products (18%) don't use any batteries (see Figure 4.9). Examples of autonomous products without batteries are: some of the PV-powered toys that were analyzed, the

1) Autonomous PV System with Batteries	2) With Battery-Chargeable	3) Without Battery	4) Hybrid

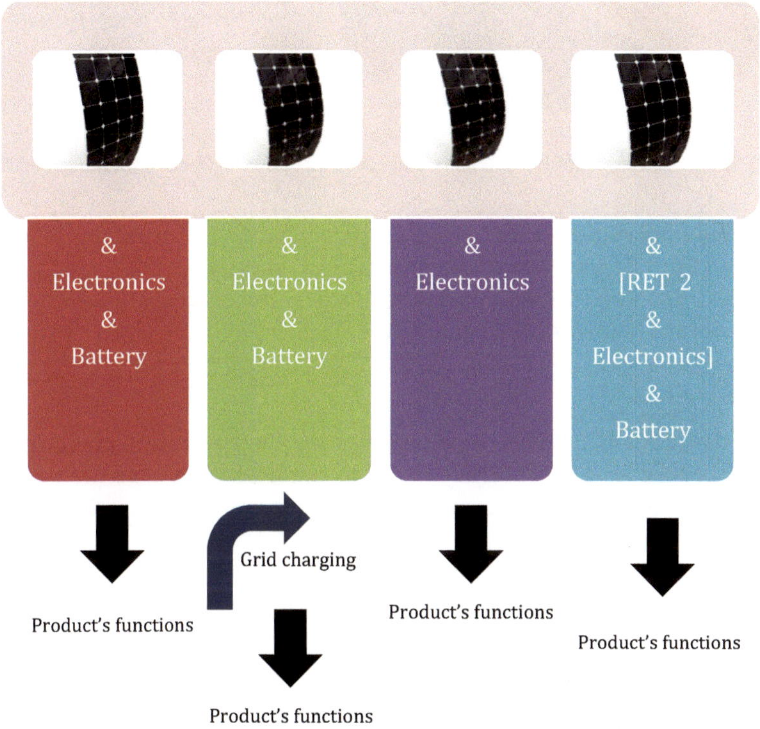

FIGURE 4.8 Schematic depiction of 4 PV system categories: (1) autonomous PV system including battery, (2) chargeable PV system including battery, (3) autonomous PV system excluding battery, and (4) autonomous hybrid PV system including battery.

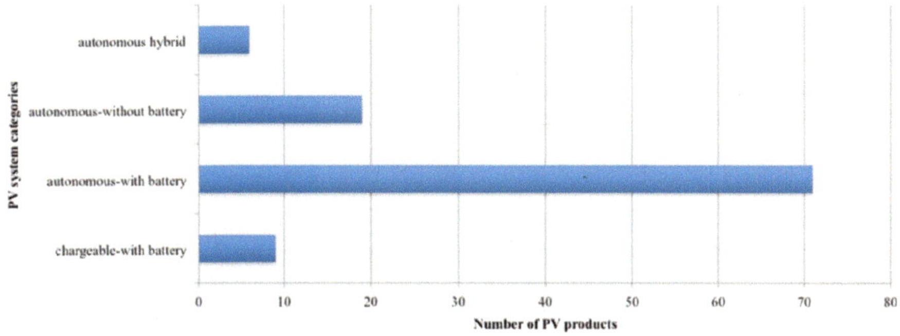

FIGURE 4.9 Amount of PV products of each PV system-category presented in Figure 4.8.

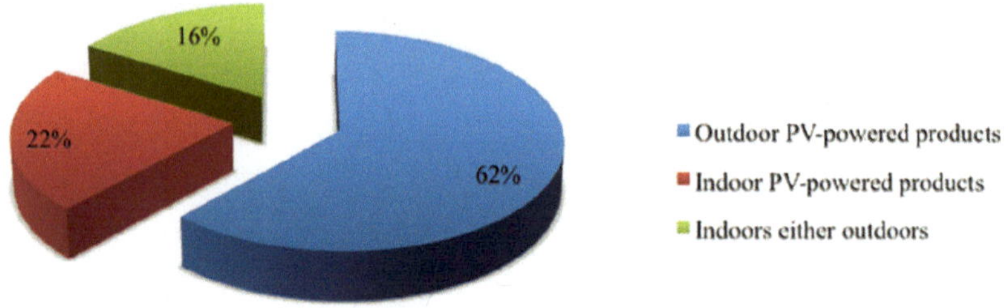

FIGURE 4.10 Share (%) of PV products that are used under different irradiance conditions.

PV-powered calculators, some PV-powered indoor lighting devices, the PV-powered sensors, and IoT devices. Approximately 62% of the analyzed PV products were mainly used outdoors, 22% were intended for indoor use, whereas 16% could be used both indoors and outdoors (see Figure 4.10).

4.3.1 PV Cells

Through the design of a PIPV, the efficiency of the PV solar cells plays a significant role because they determine, largely in combination with available irradiation, the power to be produced. The PV conversion efficiency η is defined by the ratio of the electrical power output P_{PV} (W) to the irradiance E (W/m^2) on a solar cell area A (m^2), and it is described by the following formula:

$$\eta = \frac{P_{PV}}{E \cdot A} \tag{4.1}$$

Efficiency depends on the PV technology selected, as well as the irradiance intensity, which is incident on the surface of the cell. The conversion efficiency η is determined under Standard Test Conditions (STC). Additional important factors that considerably influence efficiency are the temperature of the PV cell and the spectral distribution of the light.

In indoor environments, the measured irradiance is significantly lower than outdoors and STC. However, the performance of non-optimized cells is greatly affected by indoor irradiance. This is due to the artificial lighting systems, which are used in houses or offices and, hence, related low irradiance.

Figure 4.11 presents the types of PV technologies and the number of PV products that use each one of these technologies. It is clear that thin-film solar cells of amorphous silicon are used in most PV products examined (35.2%). Second, crystalline silicon cells are also applied to many high- and lower-power PV products (approximately 21% of the total share of the analyzed PIPV). It is worth stating that almost 2% of the analyzed PV products use organic PV cells (e.g., PV-powered toys). This appears to be a good option for the future because the use of organic PV could be safer for the environment, as well as an economical solution for low-cost products. On the other hand, high-power PV products in the range of 17 Wp to 27 kWp (15%) use amorphous silicon solar cells, whereas 20% use multi-crystalline solar cells (m-Si or x-Si) (see Figure 4.11).

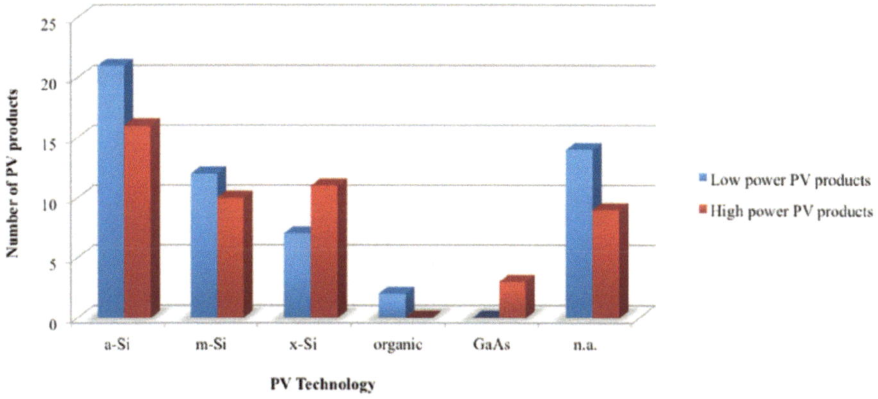

FIGURE 4.11 Number of PV products that use various types of PV technologies. Low power represents power below 17 Wp, high power above 17 Wp.

The extensive use of a-Si solar cells to numerous PV products is a consequence of their low price compared to other technologies, as well as the wide variety of sizes and shapes of a-Si cells, which are commercially accessible. The threshold for low and high power, as it is presented in Figure 4.11, is set at 17 Wp, due to the gap that was noticed among the analyzed PV products with power 17 Wp and 1 kWp. More specifically, a share of around 78% of the low-powered PV products has power between 0.01 and 10 Wp. Only 5% of the low-powered PV products have power between 10 and 17 Wp.

Regarding the high-powered PV products, it is noticed that products of that sample have power in the range of kWp. Therefore, there is a range of power between 17 and 1 kWp, where no PV products of the sample that it analyzed here belong. This is the reason that the threshold is set at 17 Wp (Apostolou and Reinders 2014).

4.3.2 Rechargeable Batteries

Batteries are used extensively in PV products in order to store electricity for use when the PV cells are not able to function properly, or to withdraw power higher than the panel cells. Approximately 82% of PIPV products have an energy storage device (see Figure 4.9). This could be a capacitor that can be used for short periods of storage (Kan 2006) or a battery, which can be used for longer periods of energy storage. Present batteries in PIPV are sulfuric lead-acid, nickel-cadmium (NiCd), nickel metal hydride (NiMH), lithium ion (Li-ion), and lithium/manganese dioxide ($LiMnO_2$) batteries. Among these technologies, nominal cell voltage can vary from 1.2 V for NiMH and NiCd batteries up to 4.1 V for Li-ion batteries. Efficiency also varies from approximately 66 % for NiMH batteries to 95%–98% for Li-ion batteries. In Table 4.2, characteristic specifications are given for cells of different rechargeable batteries applied in PIPV.

The main characteristics of batteries are their voltage, size, capacity, weight, and obviously the price. The enclosed active materials determine the battery voltage.

TABLE 4.2
Specifications of Various Rechargeable Batteries' Cells That Can Be Applied in PIPV

Battery Type	Nominal Cell Voltage (V)	Specific Energy (Wh/kg)	Energy Density (Wh/L)	Cycle Life, 20% Fading (cycles)	Specific Power (W/kg)	Efficiency[a] (%)
Lithium ion	4.1	100–265	250–730	500–1200	250–340	95–98
Lithium/ manganese dioxide	3.0	100–280	265–690	500–1000	100–315	80–95
Nickel-cadmium	1.2	40–80	50–150	1000–2000	50–200	66
Nickel metal hydride	1.2	75–120	140–300	600–1500	250–1000	80–90
Sulfuric lead-acid	2.1	30–50	80–90	500–800	75–180	

Source: Apostolou, G. and Reinders, A.H.M.E., *Energy Technol.*, 2014; Reinders, A. H. M. E. and van Sark, W. G. J. H. M., Product-integrated photovoltaics, in *Comprehensive Renewable Energy*, Vol. 1 (Ed.: A. Sayigh), Elsevier, Oxford, pp. 709–732, 2012 and updated with Intertek, 2013; Duracell, Lithium/Manganese Dioxide, http://www.duracell.com/media/en-US/pdf/gtcl/Technical_Bulletins/Lithium%20Technical%20-Bulletin.pdf; ThermoAnalytics, HEV Vehicle Battery Types, Total Thermal Solutions, http://www.thermoanalytics.com/support/publications/batterytypesdoc.html, 2013; Battery University, Lithium-based Batteries, http://batteryuniversity.com/learn/article/lithium_based_batteries, Accessed June 26, 2013.

[a] Battery efficiency: the energy efficiency of a battery is a percentage x%, which shows that x% of the energy that was put into the battery during charging is all that is available for release during discharge. The energy efficiency is given as an approximate number since discharge rates and temperature can affect it.

For example, alkaline batteries have a voltage of about 1.2 V. Lead-oxide (lead acid) batteries deliver 2.1 V, while lithium ion batteries provide 4.1 V.

From Apostolou (2016), it seems that the most-common battery technology for low-power PV products is lithium ion, while for high-power PV products lead acid or lithium batteries are preferred. Results from the analysis of the 105 PV-powered products demonstrate that 26 low-power PV products (25% of the total) use Li-ion batteries, while 6 (5.7%) use NiMH. Approximately 14% of them use no batteries. Furthermore, 23% of the low-power PV products analyzed have a battery capacity of between 1 to 3 Ah, while 20.5% have a capacity between 0.01 and 1 Ah. On the other hand, 16 high-power PV products (15%) use lead acid batteries, whereas 22 products (20%) use lithium batteries, either Li-ion or Li-poly, which are both the most common battery types that are used for these kinds of products (see Figure 4.12). Approximately 54% of the high-power PV products analyzed use batteries with a capacity between 1 to 100 Ah.

Li-ion and lead acid batteries are both the most common types of batteries that are used in products nowadays. Li-ion batteries are broadly used in product applications, due to their availability in a wide variety of shapes and sizes, suitable for the

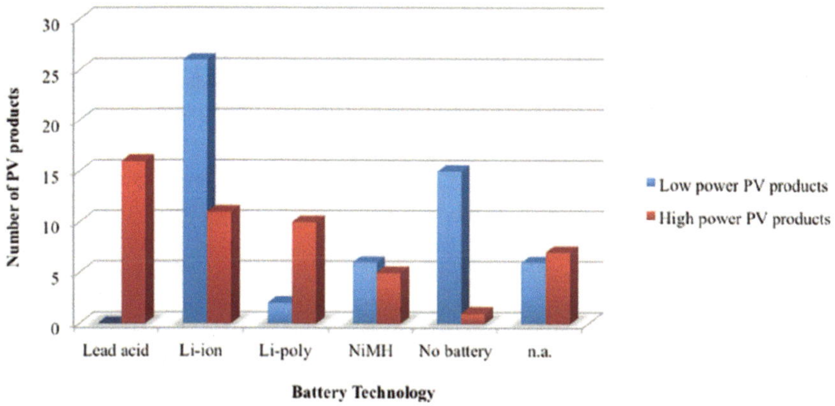

FIGURE 4.12 Number of PV products that use various battery technologies. Low power represents power below 17 Wp, high power above 17 Wp.

devices they power. Lead acid batteries are rechargeable, quite cheap, and available for purchase practically everywhere. Lead acid batteries are typically used in machinery, robotics, automobiles, and several other applications. In general, when issues such as the size or the weight of the batteries used in a product are not significant enough, and if there is a need for energy, lead acid batteries are usually preferred. Alternatively, Li-ion batteries are preferred.

Generally speaking, the analysis of Apostolou (2016) shows that there exists a correlation between PV and battery technologies. Products with power below 17 Wp use a combination of a-Si/m-Si cells and Li-ion batteries (approximately 28% of the low-power products analyzed). This is due to the low costs of these technologies, as well as their availability on the markets. On the other hand, high-power PV products (above 17 Wp) usually use a combination of Si cells (a-Si, m-Si, x-Si) and lead acid or lithium batteries (approximately 35%). The reason is that companies try to keep the cost of production as low as possible, and they choose cheap, easily available technologies for their products. Moreover, it is essential to note that recently research is turned toward the investigation of new technologies of PV cells and batteries, whose basic features are transparency and flexibility—issues that lead to a new design of products and applications.

Alternatively, in addition to the basic categories of PV technologies, research has recently focused on the production of transparent solar cells for use in buildings or products, such as e-readers and tablets (Lunt and Bulovic 2011; Chen et al. 2012; Venture Beat 2013; Zhao et al. 2014; ExtremeTech 2015; Ubiquitous Energy 2015). Research has been conducted on the fabrication of organic and polymer photovoltaic cells, which are able to absorb light in the ultraviolet and near-infrared range of wavelength and remain transparent in visible light (Lunt and Bulovic 2011; Chen et al. 2012; Zhao et al. 2014; Ubiquitous Energy 2015). PV cells with efficiencies higher than 10% and high visible transparency are available yet for incorporation in buildings, smart windows, PV products, and other PV applications. For further information regarding this kind of PV cell, you can go back to Chapter 2, where the design features of PV technologies, such as colors, flexibility, shaping, and PV cell efficiency, are discussed thoroughly. The incorporation of highly efficient PV cells in buildings is discussed in Chapter 5, where BIPVs are presented and analyzed extensively.

An alternative approach to creating transparent batteries was conducted by researchers at Stanford University (Yang et al. 2011; MIT Technology Review 2016), who were inspired by the numerous commercially available applications, such as touch screens, smartphone displays, and PV cells. It is noticeable that see-through devices have attracted considerable attention. Yang et al. (2011) succeeded in forming an entirely transparent lithium-ion battery, which was quite flexible and thin. The transparency of the electrode was a result of its feature dimension, which was below the limit of human eye' resolution. The outcome was a 60% transparent battery with 10 Wh/L energy density. In 2015, a research group from Japan at the Kogakuin University, prototyped a translucent lithium-ion (Li-ion) rechargeable battery, which charges itself using solar irradiance (Tech Xplore 2015; Nikkei Technology 2015). This battery was exhibited (in 2015) in Tokyo (Tech Xplore 2015; Nikkei Technology 2015).

Recently, in 2019, a Korean team consisting of researchers from Daegu Gyeongbuk Institute of Science and Technology, with the senior researcher Changsoon Choi, has developed transparent energy devices, using single-layered graphene film as electrodes. Results show that transparency of the energy device was increased to a maximum 77.4% (Chun et al. 2019). Additionally, a lot of research has been done on ultrathin batteries, the so-called "paper batteries" (Noyes 2007; Liat 2012; Di Wei et al. 2013, Williams 2013; Chandler 2013; Hankeun Lee and Seokheun Choi 2015; Sastikar et al. 2015). Yoon (2012) claimed that paper batteries would be soon used on bendable electronic devices, such as phones with rollable displays and e-readers.

Researchers claim that transparent batteries will be stronger soon and used in a wide range of applications (MIT Technology Review 2016; Tech Xplore, 2015). Ultrathin, flexible, stretchable, rollable, and/or batteries that could be printed, with high capacities and reliability, are extensively investigated by researchers (He 2018). Some new applications that require batteries with new form, features, and structural factors and will be soon available to the markets are: IoT applications, medical implants, wireless sensors, portable electronics, smart cards, smartphones, transportation, games, wearable textiles (He 2018).

4.4 ENVIRONMENTAL ASPECTS OF PIPV

An effective approach to quantify the environmental impacts of products is to use life-cycle analysis (LCA) during the initial phases of design and to identify the extent of the problem by imposing priorities and focusing on effective solutions. An environmental LCA is defined as: "consecutive and interlinked stages of a product system, from the raw material acquisition or the creation of natural resources to the final disposal. The main stages of a product's life include: the acquisition of raw materials, the phase before construction, manufacturing, packaging and distribution, use and end of critical life" (ISO 2006).

An LCA usually constitutes an environmental management tool, and it helps to address environmental problems through materials selection, processes of changing the product design, increased reuse, exploitation of by-products, and recycling.

"Sustainable" or "ecological" design of products establishes the incorporation of environmental features within the product design aiming to improve the conservational performance of the product throughout its lifespan. By sustainable design of PIPV products, their environmental impact should become lower than that of the current alternative designs. However, this is quite a new field of research, and it is still in progress. Therefore, the environmental aspects of products with integrated PV cells have not been addressed extensively so far. For this purpose, an LCA analysis can be used. Furthermore, other important indicators for sustainable product design are embodied energy and emissions based on common data for materials and manufacturing processes. These can be used as a rough estimation. Two examples are subsequently given concerning a detailed environmental study on solar lanterns and a more explorative one on mobile phones.

An LCA on small PV lighting products was carried out in South East Asia (Durlinger et al. 2012a, 2012b), aiming to examine the environmental impact of the production, use and end the critical life of these products. Illumination in the rural areas of developing countries is usually provided by candles and oil lamps (kerosene), whereas torches and flashes often powered by car batteries or lead-acid batteries are also used as a portable source of lighting.

Figure 4.13 and Table 4.3 present the five lighting options for developing countries that were investigated by Durlinger et al. (2012a, 2012b). These options are described extensively in the publication of Durlinger et al. (2012a, 2012b) and are the following:

- A small PV lighting system 1, which is powered by an a-Si solar panel of 0.7 Wp. It also includes two NiCd batteries of 0.2 Ah (AA-type). This product has six LED lamps of 42 lumens.

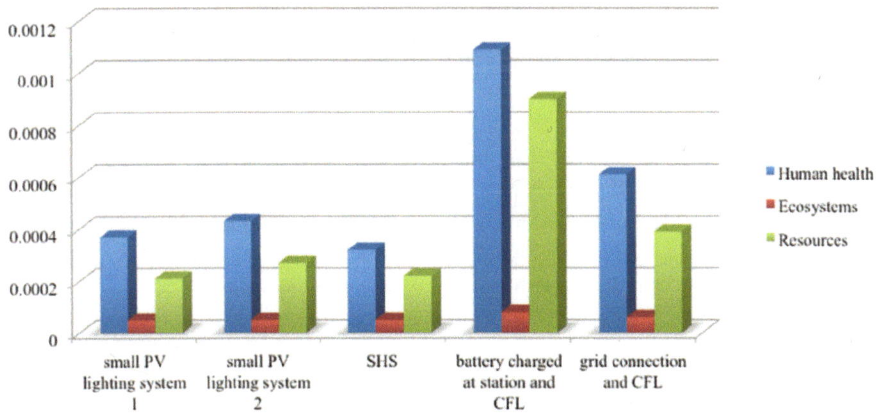

FIGURE 4.13 Results per functional unit in environmental effect groups of an LCA of small PV-powered lighting products and conventional means to provide light in rural areas in South-East Asia. The kerosene lamp is excluded. Presentation of the results on a linear scale. (Adapted from Apostolou, G. and Reinders, A.H.M.E., *Energy Technol.*, 2014.; Durlinger et al., 2012.)

TABLE 4.3

Specifications of the Five Lighting Products Evaluated by Durlinger et al. (2012)

Lighting Options for Developing Countries	Luminous Flux (lumen), Number, and Type	Battery Type, Capacity (Ah)	PV Type, Nominal Power (Wp)	Maximum Daily Operation Time (hours)
1	42, 6p, LED	NiCd, 0.2	a-Si, 0.7	3.5
2	150, 1p, 3W, CFL	Lead-acid, 4.5	x-Si, 4.5	3
3	1050, 3p, 7W, CFL	Lead-acid, 48	x-Si, 4.0	5.7
4	900, 1p, 18W, CFL	Lead-acid, 100	n.a.	60[a]
5	900, 1p, 18W, CFL	n.a.	n.a.	3

[a] Assuming that all electricity in the battery will be dissipated by lighting services.

- A small PV lighting system 2, which is powered by an x-Si solar panel of 4.5 Wp. It also includes a lead-acid battery of 4.5 Ah. This product has a compact fluorescent lamp (CFL) of 3 W and 150 lumens.
- A solar home system (SHS), with an x-Si solar panel of 40 Wp and a lead-acid battery of 48 Ah. It also includes three CFL lamps of 7 W.
- Battery charged at station and CFL. The capacity of the batteries is in the order of 100 Ah. A diesel-powered generator charges the batteries.
- Grid connection and CFL. This system includes a CFL lamp, and it is connected to the grid. There is neither solar panel, nor battery.

The study (Durlinger et al. 2012) shows that solar PV lighting products have a lower environmental effect than the conservative options for lighting in these countries have (see Figure 4.13). The data presented in Figure 4.13 are normalized (Durlinger et al. 2012). A value of 0.01 corresponds to 1%. Batteries' recycling is one way, whereby the eco-friendly profile of small-size PV lighting products can be enhanced. The authors claim that an upgrade of 10% up to 50% of the environmental profile of these products can be achieved by batteries' recycling. Moreover, it is mentioned that small-size PV lights have a lower effect to the environment, compared to grid-connected fluorescent lights. Intrinsically, PV lighting products offer an environmentally friendly and advantageous illumination service for off-grid households. Another important conclusion of this study is the possible enhancement of the environmental profile of solar lighting products by means of sufficient battery waste controlling or the use of circuit panels of a reduced size.

In Section 4.3, 19 PV lighting products are analyzed: 6 PV-powered lamps for indoor use and 13 for outdoor use. The low-powered PV lighting products with power in a range of a few Wp (0.1–10 Wp) are mainly designed to be charged outdoors and used indoors. These products have a small PV area, usually made by either mono- or multi-crystalline silicon cells, which could be integrated in the lamp's surface or consist a separate product connected to the lamp. The area of the PV cell depends on the available surface on the product and is usually a few cm^2. PV-powered lights also include one or more small rechargeable batteries with capacity 1.2–4.1 Ah each one, depending on their type (e.g., lithium ion, NiMH, NiCd). The limited area of the PV cell, as well as the rechargeable batteries, which can be recycled, offer an environmentally friendly character to these lighting products, since these elements contribute to a safer waste controlling than the grid-connected fluorescent lamps. However, a further analysis of the environmental impact of these products is out of the scope of this chapter and is not further detailed.

Another study provides a battery life-cycle assessment for mobile phones with a focus on cradle-to-grave (CTG) energy and greenhouse gas (GHG) emissions (Flipsen et al. 2012; Dafnomilis 2012). The study is based on bibliographical data and on a comparison of four different types of smartphones using an original battery and after a reduction of the battery capacity while using a solar cover (see Figure 4.14b). The solar cover that was used for the tests and estimations belongs to the mobile phone GD510 by LG (see Figure 4.14a and b). GD510 was introduced as a solar-powered phone. It includes an additional back cover with an embedded solar panel. It has the regular option to be charged through the electricity grid, but it can also be charger by solar irradiation. According to the manufacturer's specifications, with 10 minutes of charging under direct sunlight, the user can have a battery lifetime of 3 hours when the phone is idle, or 2.5 hours of talk time.

The four smartphones, as presented in Figures 4.15 through 4.17 and their battery specifications, are the following:

- Samsung Galaxy SII (Li-ion battery, 3.7 V, 6105 mWh)
- Sony Ericson Xperia Mini (Li-polymer, 3.7 V, 4440 mWh)
- HTC Wildfire (Li-polymer, 3.7 V, 4810 mWh)
- Blackberry Torch 9800 (Li-ion, 3.7 V, 4700 mWh)

The modifications made to these smartphones include an x-Si solar cover (like the one presented in Figure 4.14b) and a new battery with approximately 30% lower capacity

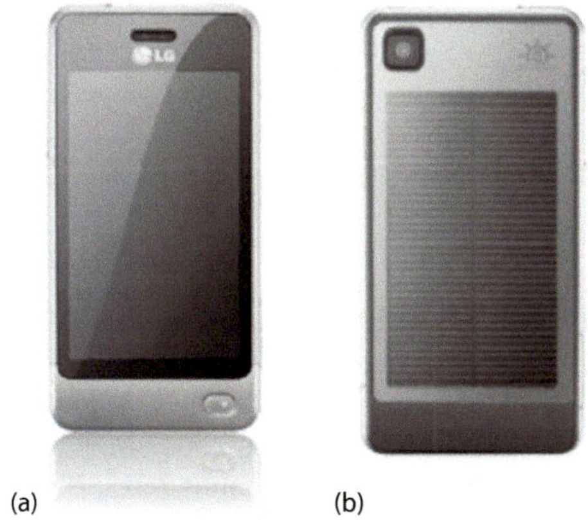

FIGURE 4.14 (a) The GD510 mobile phone by LG. It was introduced as a solar-powered phone, and (b) the phone includes an additional back cover with an embedded solar panel. (Courtesy of LG.)

FIGURE 4.15 Smartphones used during the study of the battery life-cycle assessment: (a) Samsung Galaxy SII, (b) Sony Ericson Xperia Mini, (c) HTC Wildfire, and (d) Blackberry Torch 9800.

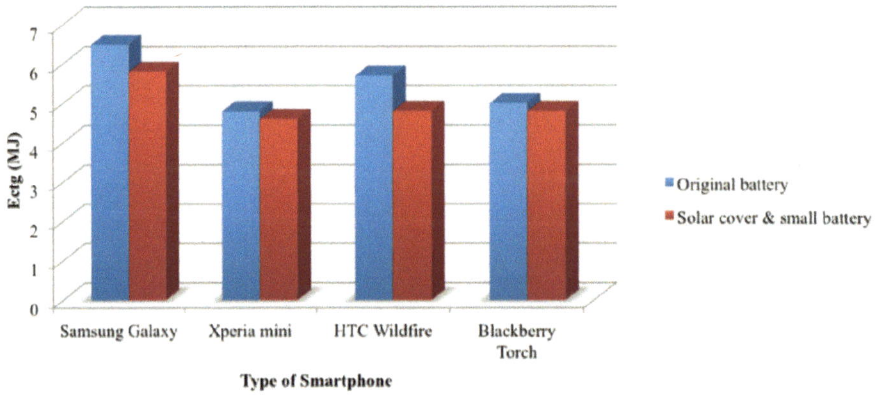

FIGURE 4.16 Comparison of the total energy (Ectg) required during the life cycle of four mobile phones between two different technologies; original battery and solar PV cover with a smaller battery. (Adapted from Apostolou, G. and Reinders, A.H.M.E., *Energy Technol.*, 2014; Dafnomilis, 2012.)

than the original battery of each smartphone (Dafnomilis 2012; Flipsen et al. 2012). The typical back area of an average smartphone is calculated around 0.0067 m^2, and 70% of this area can be covered with a PV panel. This means that the area of the smartphones' solar back covers will be around 0.0047 m^2 with power 72 mW (using solar energy data for The Netherlands) (Dafnomilis 2012; Flipsen et al. 2012).

Findings from this study (Flipsen et al. 2012; Dafnomilis 2012) indicate that a reduction of 30% of the batteries' capacities will result in a reduction of approximately 30% of the energy consumed during the production of raw materials, manufacturing, and CO_2 emissions during the battery's lifetime. From this point of view,

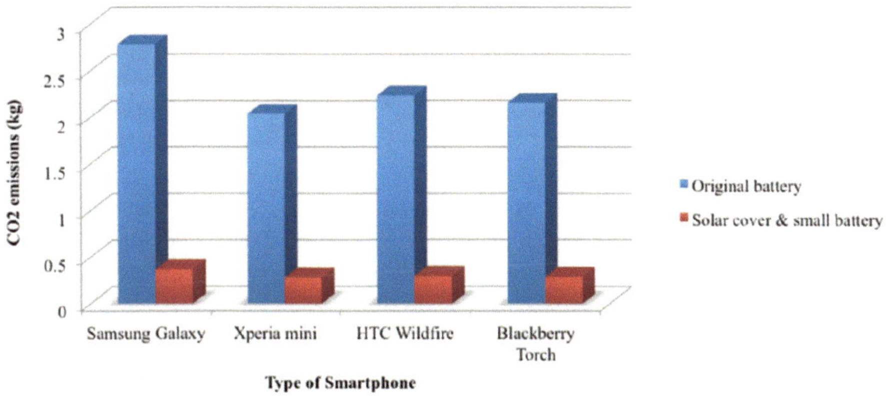

FIGURE 4.17 Comparison of the amount of CO_2 emissions required during the life cycle of four smartphones, between two different technologies; original battery and solar PV cover with a smaller battery. (Adapted from Apostolou, G. and Reinders, A.H.M.E., *Energy Technol.*, 2014; Dafnomilis, 2012.)

solar phones can be an environmentally friendly solution to oversized batteries—especially for moderate and light users or in countries with sufficient annual sunshine. Besides, they can be an ideal solution for off-grid areas. Results from this study are presented in Figure 4.10, where the total energy comparison between two different technologies—a mobile phone without modification and one with a smaller battery including a solar cover—for four types of smartphones is indicated. In Figure 4.11, a comparison of CO_2 emissions is presented, respectively.

As Figures 4.16 and 4.17 illustrate the use of a solar back cover combined with a battery of reduced capacity that offers benefits to the energy that is consumed during the extraction process of raw materials, the manufacturing process of the products, and to the CO_2 emissions during these procedures. CO_2 emissions required during the life cycle of the original smartphones are significantly higher than CO_2 emissions required for the modified smartphones (see Figure 4.17). The solar back cover usually has a longer lifetime than the batteries. This means that when the battery is at the end of its life, the solar cover is not necessary to be changed. The solar cover could be used with a new battery on the same smartphone or else could be modified and integrated to another smartphone. This might be an advantageous opportunity for the foreseeable future; the use of product's assembling parts to other products. In that way, materials are recycled and products' environmental profiles are enhanced.

4.5 HUMAN FACTORS OF PIPV

In the stage of the implementation of PIPV products, users' involvement and awareness of the products are significant. Consumers prefer to buy products that not only have sufficient functions but also because of their looks. They desire products with a nice visual appeal, color, material, and design, as well as products that raise their emotional reactions.

Although human factors constitute a key element for the successful transaction of a product, there is actually little research published on this subject with regard to PIPV. On the other hand, studies regarding users' interactions with products have been conducted toward sustainable design (co-design) by many researchers (Bakker et al. 2010; Jong and Maze 2010; Keyson and Jin 2009; Wever et al. 2008; Scott 2008; Scott et al. 2009; Verbeek and Slob 2006; Jong et al. 2008). Regarding user's interactions with PIPV products only one extensive study is available. This study concerns the assessment of the Sole Mio, a prototype of a PC computer mouse (Figures 4.3g and 4.4). It was carried out during the period September–December 2007 in the Netherlands (Reich et al. 2008). In this study, 14 people participated and asked to use the Sole Mio mouse daily for some weeks. A number of users had a test of 5–6 weeks, while others an extended one of 10–12 weeks.

Part of the results of the user tests focused on the general user expectations, experiences with charging batteries, reactions to the feedback signal for battery charging, and user willingness to buy a Sole Mio (Reich et al. 2008). The outcomes of the study are summarized below:

The majority of the users, at the beginning of the test procedure, were completely unaware of the test's expectations. The performance of the mouse varied significantly from one day to another, and this generated a feeling of doubt about and unreliability of the product. Generally speaking, users' impressions concerning the Sole Mio mouse was negative. Furthermore, the period when the test took place was during the wintertime, when the solar irradiation was considerably lower than during the other periods of the year. This was not only a difficult undertaking for the performance of the product, but it was also an interesting experiment concerning the worst-case scenario of usage. If the mouse operates sufficiently during this period, then it will perform even better during the lighter days of the year.

Regarding charging the battery of the Sole Mio, the solar cell in the product was exposed to a higher irradiance on the windowsill, through so-called sunbathing. Each user followed its private charging strategy. Some users charged the mouse the entire day on the windowsill, while others were charged it once per week, whenever the mouse's battery was empty. The evaluation's outcome was diverse. Undoubtedly, the mouse had a satisfactory performance when treated well by charging when required. Moreover, the mouse included a light sign, which indicated that charging is in process or that the mouse needs to be charged. Nonetheless, many users claimed that the existence and the function of the light indicator were rather vague to them.

Based on the user tests and interviews previously mentioned, it is clear that some users were more excited about the Sole Mio mouse than others. Satisfied users might be willing to spend around €50 to buy the product. However, since the mouse is still not available on the market, despite being technically feasible yet, it is questionable whether users would still be willing to buy it or not. Unfortunately, this issue cannot be tested.

Finally, the experiment pinpointed two categories of customer. The first category includes those who were enthusiastic about the Sole Mio mouse and believed that it could be a good choice. However, they were not willing to buy it because they would prefer a more reliable and low-priced mouse instead, such as the normal wired mouse. The second category consists of customers who would be willing to buy the

Sole Mio mouse. However, they only agreed to offer about €10 more for the Sole Mio mouse than for a conventional mouse. If the cost were higher, it is doubtful whether they would actually buy it.

4.6 COSTS OF PIPV

By the previous findings regarding the Sole Mio mouse, the topic of costs of PIPV has been introduced. Cost of a product is the expense incurred by a corporation in order to sell it. It might include raw, material, production, labor, tooling, utilities, operating, shipping costs, and other possible expenditure until the product will be finally commercially available. Therefore, it is quite difficult to accurately estimate the costs of a product without access to the company's data (e.g., product manufacturers, product development departments).

However, even if the cost of a product is difficult to be predicted, the sales price is easily obtainable and it is the one, which directly concerns the consumer. Industries target on low-cost products, which are affordable to consumers. In order to succeed in low sales prices, companies usually choose low-cost materials and services during the manufacturing process of the product. On the one hand, this could lead to cheap products and high sales, but on the other hand to low quality products with an ambiguous performance.

Concerning the sales price of PV products, the results of the analysis presented here show that there is a wide range of prices either for indoor or outdoor PV products (Figure 4.18). Outdoor products, such as solar boats, aircrafts, or cars, are quite expensive, and some of them cost several thousands of Euros. This is due to their complex production, size, the cost of materials, and the human effort that is needed and the time of manufacturing. In general, most of the products, which belong to the categories of recreational PV products, solar vehicles and transportations, and business-to-business PV applications, as well as art PV products, have high sale prices ranging

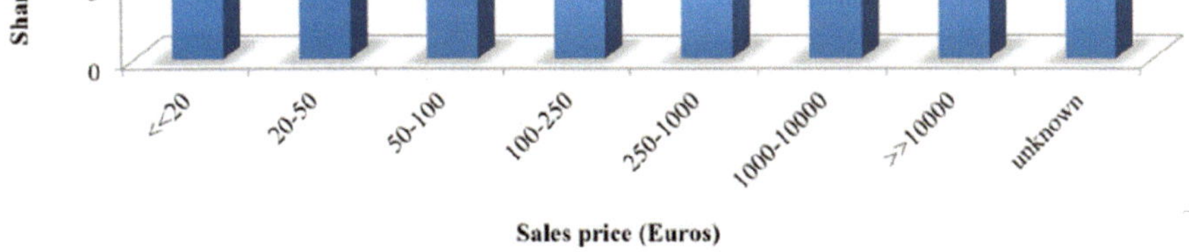

FIGURE 4.18 Sale prices in Euro of numerous PV products both for indoor and outdoor use. (Adapted from Apostolou, G. and Reinders, A.H.M.E., *Energy Technol.*, 2014 and updated for 2019 by Apostolou.)

from hundreds to thousands of Euros (see Figure 4.18). However, there are multiple outdoor PV products that are quite affordable, such as solar lighting products or consumer products. The sales price of these products ranges between 10 Euros (e.g., solar garden lights) and 100 Euros (e.g., solar lawn mowers) (Apostolou and Reinders 2012).

Indoor PV products usually have lower costs than outdoor do. However, this is not a rule because there are also some rather expensive indoor PV products (e.g., i-Phone charger) (Figures 4.1c and 4.2). Several products such as indoor PV lights, solar toys, solar desk lamps, and PV chargers might cost few Euros (from less than 10 Euros to 50 Euros), as Figure 4.18 illustrates. The sales price of these products is defined by the design complexity, materials, originality of the idea, the concept of the product, and many other factors. Over the following years, it is expected that the cost of product manufacturing, as well as product sales price will be reduced, as new technologies are investigated in PV materials and batteries with higher performances, better technical features, and lower costs.

4.7 CONCLUSIONS

This chapter presented several categories of PIPVs and addressed various design features of PV products. PIPV can be applied well in different product categories of various power levels, ranging from several mWs to several kWs. Surprisingly, art is a new field in which PV has recently been introduced because high-energy supplies are not required for art products to work, while, at the same time, nice visual appeal is sought.

Four PV system categories were distinguished. About 71 out of 105 PV products analyzed consist of an autonomous PV system with batteries. Approximately 62% of the analyzed PV products are used outdoors, while around 22% are used indoors and 16% both indoor and outdoor. Approximately 37.5% of the low-power PV products in the range of 0 to 17 Wp use thin-film solar cells (a-Si), whereas 75.5% of high-power PV products in the range of 17 to 27 kWp use x-Si, m-Si, or a-Si solar cells. Approximately 82% of PIPV products use an energy storage device, while 18% do not use any batteries.

The environmental impact of PIPV products was explored using the example of smartphones and a PV lighting product in South East Asia. Human factors were also addressed, using the example of the PC computer mouse Sole Mio. On the basis of these cases, it was concluded that the environmental impact could be reduced by recycling batteries, replacing original batteries with smaller ones, or using an additional solar cover. However, this is not feasible for all products or for all user types. On the other hand, concerning the environmental impact of PV materials, studies by Durlinger (2010, 2012) claim that solar lighting systems have a considerably lower environmental effect than the conservative lighting options, which are broadly used in off-grid areas.

The example of the computer mouse Sole Mio shows that the performance of the product depends mainly on the user's behavior. In this example, it was revealed that the performance of the mouse varies according to user charging tactics. Some users were willing to sunbathe the mouse for a longer period or more often than others. As a result, their mouse performed better than those that were charged infrequently.

While designing products with integrated PV, the focus should be on both the environmental impact and human factors. These two aspects are closely related to each other and should be analyzed together. Despite their importance for the final product experience, they have not been evaluated extensively so far.

Based on our evaluation of existing knowledge, it is believed that PIPV will be further developed in the years to come. Likewise, it is expected that new PV products will be launched for both outdoor and indoor use, such as sensor networks, indoor lighting, luminescent solar concentrator photovoltaic (LSC-PV) street lighting (Viswanathan et al. 2012), furniture, electric vehicles, or art products. Furthermore, new PV technologies with enhanced design features, such as coloring, flexibility, and transparency, are entering the PIPV market (see also Chapter 2).

REFERENCES

Alsema, E. A., B. Elzen, N. H. Reich, W. G. J. H. M. van Sark, S. Y. Kan, S. Silvester, M. Veefkind, J. Jelsma, Towards an optimized design method for PV-powered consumer and professional applications—the Synenergy project. In *Proceedings of the 20th European Photovoltaic Solar Energy Conference, Barcelona* 2005 (Eds.: W. Palz, H. Ossenbrink, P. Helm), Spain, WIP-Renewable Energies, 1981–1984, 2005.

Anweshas Solar Powered LED Wall Light with PIR Motion Sensor/CDS Night Sensor Lamp Pack. https://www.clubfactory.com/amp/item-PID-3461455.html. Accessed July 1, 2019.

Apostolou, G., Design Features of Product-Integrated PV: An Evaluation of Various Factors under Indoor Irradiance Conditions, PhD Thesis, Industrial Design Engineering Department, Technical University of Delft (TU Delft), Delft, the Netherlands, 2016.

Apostolou, G., A. H. M. E. Reinders, A comparison of Design Features of 80 PV-Powered Products, *Proceedings of the 27th EUPVSEC, Poster Presentation*, Frankfurt, Germany, 2012, pp. 4227–4232.

Apostolou, G., A. H. M. E. Reinders, Overview of design issues in product- integrated photovoltaics, *Energy Technology* 2014. doi:10.1002/ente.201300158.

Bakker, C. A., D. J. van Eijk, S. Silvester, M. E. F. R. Reitenbach, A. M. de Jong, D. V. Keyson, K. Scott. (2010). Understanding and modelling user behaviour in relation to sustainable innovations: the living lab method, In *Proceedings of TMCE 2010 Symposium* (Eds.: I. Horvath, F. Mandorli, Z. Rusak) April 12–16, Ancona 2010, Italy, pp. 1-12, s.l.: TMCE (TUD).

Battery University, Lithium-based Batteries. http://batteryuniversity.com/learn/article/lithium_based_batteries. Accessed June 26, 2013.

Blackberry Torch: https://www.google.com/search?q=Blackberry+Torch+2014&source=lnms&tbm=isch&sa=X&ved=0ahUKEwiav9G02bbiAhWSY1AKHWNPAB8Q_AUIDigB&biw=1280&bih=611#imgrc=8Sr72J1gdyThtM. Accessed July, 1, 2019.

Bringing the butterfly effect to life, PV, Renewable Energy Magazine: http://www.renewableenergymagazine.com/article/bringing-the- butterfly-effect-to-life. Accessed February 13, 2014.

Brown, E. (September 13, 2016a). Who Needs the Internet of Things? Linux.com. Accessed May 25, 2019.

Brown, E. (September 20, 2016b). 21 Open Source Projects for IoT. Linux.com. Accessed May 25, 2019.

Business Insider. (2017). How IoT & smart home automation will change the way we live. Accessed May 25, 2019.

Business Insider Intelligence. (2020). How IoT devices and smart home automation is entering our homes in 2020, https://www.businessinsider.com/iot-smart-home-automation. Accessed February 10, 2020.

Chandler, D. L., Research update: Team observes real-time charging of a lithium-air battery, *MIT News*. http://web.mit.edu/newsoffice/2013/real-time-charging-of-lithium-air-battery-0513.html, 2013.

Chen, C. C., L. Dou, R. Zhu, C. H. Chung, T. B. Song, Y. B. Zheng, S. Hawks, G. Li, P. S. Weiss, Y. Yang, *ACS Nano* 2012, 6, 7185–7190. doi: 10.1021/nn3029327.

Chun, S., Son, W., Lee, G. et al. Single-layer graphene-based transparent and flexible multifunctional electronics for self-charging power and touch-sensing systems. *ACS Applied Materials & Interfaces* 11(9), 9301–9308. doi:10.1021/acsami.8b20143.

Dafnomilis, I., Battery capacity and performance in relation to user profiling in smartphones, Student report on System Integration Project II (SIP 2), TU Delft, the Netherlands, 2012.

Duracell, Lithium/Manganese Dioxide. http://www.duracell.com/media/en-US/pdf/gtcl/Technical_Bulletins/Lithium%20Technical%20- Bulletin.pdf.

Durlinger, B., A. Reinders, M. Toxopeus, Environmental benefits of PV powered lighting products for rural areas in South East Asia: A life cycle analysis with geographic allocation. In: *Proceedings of 35th IEEE PVSC*, Honolulu, HI, 2010, pp. 2353–2357.

Durlinger, B., A. Reinders, M. Toxopeus, Life cycle assessment of solar powered lighting products for rural areas in South East Asia, Components for PV Systems, PV System Engineering and Standards, Socio-economic Aspects, and Sustainability: *Proceedings of 25th EU PVSEC*, Valencia, Spain, 2012a, pp. 3917–3925.

Durlinger, B., A. Reinders, M. Toxopeus, A comparative life cycle analysis of low power PV lighting products for rural areas in South East Asia, *Renewable Energy* 2012b, 41, 96–104. doi:10.1016/j.renene.2011.10.006.

ExtremeTech. (2015). This fully transparent solar cell could make every window and screen a power source (updated), By Jamie Lendino on April 20, 2015. http://www.extremetech.com/extreme/188667-a-fully-transparent-solar-cellthat-could-make-every-window-and-screen-a-power-source.

Flipsen, B., J. Geraedts, A. Reinders, C. Bakker, I. Dafnomilis, A. Gudadhe, Environmental sizing of smartphone batteries, *Proceedings Micro-Energy Supplies and Energy Harvesting: Electronics Goes Green* 2012, Berlin, Germany, 2012, pp. 1–9.

Freunek, M., M. Freunek, L. M. Reindl, Maximum efficiencies of indoor photovoltaic devices, *IEEE Journal of Photovoltaics* 2013, 3, 59–64.

Hçsel, M., R. R. Sondergaard, M. Jørgensen, F. C. Krebs, Fast inline roll-to-roll printing for indium-tin-oxide free polymer solar cells using automatic registration, *Energy Technology* 2013, 1, 102–107.

He, X., Flexible, Printed and Thin Film Batteries 2019–2029, 2018. https://www.idtechex.com/de/research-report/flexible-printed-and-thin-film-batteries-2019-2029/634. Accessed June 28, 2019.

Hendricks, D. The Trouble with the Internet of Things. London Datastore. Greater London Authority. Accessed May 25, 2019.

HTC Wildfire S: https://www.google.com/search?biw=1280&bih=611&tbm=isch&sa=1&ei=4zbpXOTcGuPjgweNvbI4&q=HTC+Wildfire+2014&oq=HTC+Wildfire+2014&gs_l=img.3...11392.12707..13370...0.0..0.129.591.0j5......0....1..gws-wiz-img.......0i19j0i30i19.i2prFfTx-gM#imgdii=UU1-6OqnqshuyM:&imgrc=EjoTP0CJqC_ooM. Accessed July, 1, 2019.

ITU. (2015) Internet of Things Global Standards Initiative. https://www.itu.int/ITU-T/recommendations/rec.aspx?rec=y.2060. Accessed May 25, 2019.

Jong, A. M., C. A. Bakker, K. Scott, The Living Lab project; user centered sustainable design, Sustainable Innovation 2008, Delft University of Technology, Wuppertal Institute, Swiss Federal Institute of Technology in Zurich, Technical University of Madrid, BASF, Acciona, Proctor & Gamble, Cincinnati, OH, 2008.

Jong, A., R. Mazé. Cultures of sustainability-ways of doing cooking, 14th European Roundtable on Sustainable Production and Consumption conference (ERSCP), 6th Environmental Management for Sustainable Universities conference (EMSU), Knowledge Collaboration & Learning for Sustainable Innovation, October 25–29, Delft, The Netherlands, 2010.

Kan, S. Y., SYN-Energy in solar cell use for consumer products and indoor applications, PhD diss., Final Report 014–28–213, NWO/NOVEM, Technical University of Delft, Delft, the Netherlands, 2006.

Kan, S. Y., M. Verwaal, H. Broekhuizen, The use of battery-capacitor combinations in photovoltaic powered products, *Journal of Power Sources* 2006, 162, 971–974.

Kang, W. M., S. Y. Moon, J. H. Park, An enhanced security framework for home appliances in smart home. *Human-centric Computing and Information Sciences* 2017, 7 (6). doi:10.1186/s13673-017-0087-4.

Keyson, D., S. Jin, (Eds.) *Designing for Sustainable Living & Working*, Delft, The Netherlands: VSSD, 2009.

Laser Focus World, Organic PV efficiency reaches 12%: Heliatek breaks its prior record, International Resource for Technology and Applications in the Global Photonics Industry. http://www.laserfocus-world.com/articles/2013/01/Heliatek-OPV-12-percent.html, 2013.

Lee, H., Choi, S. An origami paper-based bacteriapowered battery, *Nano Energy*, 2015, 15, 549–557.

Lee, J.-H., T. Sagawa, S. Yoshikawa, Fast screening of the optimal polymer ratio for organic solar cells using a spray-coating deposition method for fullerene mixture, *Energy Technology* 2013, 1, 85–93.

LG: https://www.lg.com, https://www.treehugger.com/clean-technology/lg-pop-cell-phone-goes-solar-powered.html. Accessed July 1, 2019.

Liat, C., Ultra-thin flexible lithium battery keeps powering when bent, WIRED.CO.UK, http://www.wired.co.uk/news/archive/2012–08/08/ bendy-battery, 2012.

Lunt, R.R., V. Bulovic. Transparent, near-infrared organic photovoltaic solar cells for window and energy-scavenging applications, *Applied Physics Letters*, 2011, 98, doi:10.1063/1.3567516.

MIT Technology Review, Startup Engineers See-Through Solar Cells. http://www.technologyreview.com/news/511681/startup-engineers-see-through-solar-cells/, 2016.

Nanowerk, Nanotechnology enhanced organic photovoltaics: Breaking the 10% efficiency barrier. http://www.nanowerk.com/spotlight/ spotid=29503.php, 2013.

Netcomposites, Helios Solar-Powered Aircraft Crashes, NASA.gov: http://www.netcomposites.com/news/helios-solar-powered-aircraft- crashes/1736, Accessed February 13, 2014.

Nikkei Technology, Translucent Li-ion Battery Charges Itself by Using Sunlight, September 1, 2015, http://techon.nikkeibp.co.jp/atclen/news_en/15mk/090100006/?n_ cid=nbptec_tecrs, 2015.

Noyes, K., Nanotubes Power Paper-Thin Battery, Tech News World. http://www.technewsworld.com/story/58833.html, 2007.

"Nuna" Solar-Powered Car, Delft University of Technology, The Netherlands, Hans-Peter van Velthoven: http://www.hanspeter.nl/ #/nuna/20. Accessed February 13, 2014.

O'Hara, T., M. Wesselmark, Battery Technologies: A General Overview & Focus on Lithium-Ion, Intertek booklet available at http://www.kta-ev.com/v/vspfiles/assets/images/intertek-battery-technologies.pdf, 2012.

PlanetSolar Turanor, BlueBird-Electric.net: http://www.bluebird-electric.net/planetsolar.htm. Accessed February 13, 2014.

PowerFilm Solar, Bluetooth Indoor Solar Development Kit (LES100). https://www.powerfilmsolar.com/how-to-buy/shop-online/bluetooth-indoor-solar-development-kit. Accessed July 1, 2019.

Randall, J. F., Designing Indoor Solar Products: Photovoltaic Technologies for AES, PhD diss., ETHZ Switzerland, John Wiley & Sons, 2005.

Reich, N. H., B. Elzen, M. P. Netten, W. G. J. H. M. van Sark, E. A. Alsema, M. Veefkind, S. Silvester, Practical experiences with the PV powered computer mouse "Sole Mio," Components for PV Systems in Engineering and System Integration: *Proceedings of 23rd European Photovoltaic Solar Energy Conference*, Valencia, Spain, 2008, pp. 3121–3125.

Reich, N. H., M. Veefkind, W. G. H. M. van Sark, E. A. Alsema, E. A. Turkenburg, S. Silvester, A solar powered wireless computer mouse: Industrial design concepts, *Solar Energy* 2009, 83, 202–210.

Reich, N. H., M. P. Netten, M. Veefkind, W. G. J. H. M. van Sark, E. A. Alsema, S. Silvester, B. Elzen, M. Verwaal, A solar powered wireless computer mouse: Design, assembly and preliminary testing of 15 prototypes: *Proceedings of 22nd European Photovoltaic Solar Energy Conference*, Milan, Italy, 2007.

Reich, N. H., W. G. J. H. M. van Sark, E. A. Alsema, H. de Wit, A. H. M. E. Reinders, A CAD based simulation tool to estimate energy balances of device integrated PV systems under indoor irradiation conditions, *Proceedings of 23rd European Photovoltaic Solar Energy Conference, PV Systems, Off-grid Applications*, Valencia, Spain, 2008, pp. 3338–3343.

Reinders, A. H. M. E., Options for photovoltaic solar energy systems in portable products, in *Proceedings of the TMCE*, Wuhan, PR China, 2002.

Reinders, A. H. M. E., G. Apostolou, Product integrated PV, Book Chapter in *Photovoltaic Solar Energy: From Fundamentals to Applications* (Eds.: A. H. M. E. Reinders, P. Verlinden, W. G. J. H. M. van Sark, A. Freundlich), John Wiley and Sons, London, UK, January 2017.

Reinders, A. H. M. E., J. C. Diehl, H. Brezet, The Power of Design: Product Innovation in Sustainable Energy Technologies, Wiley, Delft, 2012.

Reinders, A. H. M. E., W. G. J. H. M. van Sark, Product-Integrated Photovoltaics: in *Comprehensive Renewable Energy*, Vol. 1 (Ed.: A. Sayigh), Elsevier, Oxford, 2012, pp. 709–732.

Samsung Galaxy SII: https://gadgets.ndtv.com/samsung-galaxy-s-ii-554. Accessed July 1, 2019.

Sastikar Anushri, S., S. Bobade Trupti, T. Sneha, Carbon nanotube based paper battery and lithium ion battery, *International Journal of Application or Innovation in Engineering & Management (IJAIEM)*, 2015, 4, 178–185.

Scott, K., Co-designing sustainable user practices, Thesis in Industrial Ecology master program, TU Delft, Erasmus University, Leiden University, Leiden, the Netherlands, 2008.

Scott, K., J. Quist, C. Bakker, Co-design, social practices and sustainable innovation: involving users in a living lab exploratory study on bathing, Joint actions on climate change conference, June 8–10, Aalborg, Denmark, 2009.

Smart 2.0, Energy harvesting PMIC targets solar-powered wireless sensors. https://www.smart2zero.com/news/energy-harvesting-pmic-targets-solar-powered-wireless-sensors. Accessed July 1, 2019.

Solar Bonsai Electree, Ecological charger for mobile devices by Vivien Muller: https://mu-design.lu/mu-design-en#electree-original. Accessed June 28, 2019.

Solar-powered Electree mini by Vivien Muller: https://mu-design.lu/mu-design-en#electreemini. Accessed June 28, 2019.

Solar-powered Orkys by Vivien Mueller: https://mu-design.lu/mu-design-en#orkys. Accessed June 28, 2019.

Solar Powered Mouse—The Sole Mio, photo by Matthijs Netten, Delft Design Institute/SynEnergy: http://www.tudelft.nl/en/current/ latest-news/article/detail/minister-cramer-test-eerste-muis-op- zonnecellen/. Accessed February 13, 2014).

Sony Ericsson Xperia mini: https://www.amazon.fr/Sony-Ericsson-Smartphone-Appareil-Allemagne/dp/B0051XZIMG. Accessed July 1, 2019.

Tech Xplore, Transparent lithium-ion battery that recharges via the sun demonstrated, September 2, 2015 by Bob Yirka, http://techxplore.com/ news/2015-09-transparent-lithium-ion-battery-recharges-sun.html, 2015.

TEKBOX Digital Solutions, Solar panel powered LoRaWAN RTU for SDI-12 sensors—TBS12S. https://www.sdi-12products.com/products/solar-panel-powered-lorawan-rtu-for-sdi-12-sensors-tbs12s. Accessed July 1, 2019.

ThermoAnalytics, HEV Vehicle Battery Types, Total Thermal Solutions. http://www.thermoanalytics.com/support/publications/ batterytypesdoc.html, 2013.

Timmerman, M. B., Review of Existing Knowledge of Product Integrated PV for Industrial Design Engineers, Internal Research Paper, Master of Sustainable Energy Technology, University of Twente, 2008.

Ubiquitous Energy, ClearView Power Technology, http://ubiquitous.energy/technology/, 2015.

User Behavior and Technology Development: Shaping Sustainable Relations Between Consumers and Technologies (Eds.: P.-P. Verbeek, A. Slob), Vol. 20, Springer, Heidelberg, Germany, 2006.

Veefkind, M., N. H. Reich, B. Elzen, S. Y. Kan, S. Silvester, M. Verwaal, E. A. Alsema, W. G. J. H. M. van Sark, J. Jelsma, The design of a solar powered consumer product, a case study: in *Proceedings of Going Green—Care Innovation 2006, International Journal of Automation*, Vienna, Austria, 2006, pp. 3.6.4.

Venture Beat VB/GREEN, These transparent solar cells provide 50% more battery life for your phone-and infinite standby time. http://venturebeat.com/2013/06/04/these-transparent-solar-cells-provide-50-more-battery-life-for-your-smartphone-and-give-you-infinite-standby-time/, 2013.

Verbeek, P.-P., A. Slob (Eds.) *User Behavior and Technology Development, Eco-Efficiency in Industry and Science*, 2006, Vol. 20, XVI, 409 p., ISBN 978-1-4020-5196-8.

Viswanathan, B., A. H. M. E. Reinders, D. K. G. de Boer, A. Ras, H. Zahn, L. Desmet, System engineering and design of LSC-PV for outdoor lighting applications: *Proceedings of EUPVSEC*, Frankfurt, Germany, 2012, pp. 4273–4279.

Vivien Mueller, iPhone Chargers Electree, https://mu-design.lu/mu-design-en#electree-original, © Vivien Mueller, Accessed July 1, 2019. Vivien Mueller, Electree Mini, https://mu-design.lu/mu-design-en#electreemini, © Vivien Mueller, Accessed July 1, 2019.

Wei, D., Haque, S., Andrew, P., Kivioja, J., Ryhänen, T., Pesquera, A., Centeno, A., Alonso, B., Chuvilincd, A., Zurutuzab, A., Ultrathin rechargeable all-solid-state batteries based on monolayer grapheme, *Journal of Materials Chemistry A*, 2013, 1, 3177–3181. doi: 10.1039/C3TA01183F.

Wever, R., van J. Kuijk, C. Boks, User-centred design for sustainable behaviour, *International Journal of Sustainable Engineering*, 2008, 1 (1), 9–20.

Williams, M., Hybrid ribbons a gift for powerful batteries, Rice University News & Media. http://news.rice.edu/2013/03/25/hybrid-rib- bons-a-gift-for-powerful-batteries/, 2013.

World Solar Challenge. https://www.worldsolarchallenge.org/, Accessed February, 2020.

Yang, Y., S. Jeong, L. Hu, H. Wu, S. W. Lee, Y. Cui, Transparent lithium-ion batteries: in *Proc. Natl. Acad. Sci. USA*. doi:10.1073/ pnas.1102873108, 2011.

Yoon, L., A KAIST research team has developed a high performance flexible solid-state battery, EurekAlert. http://www.eurekalert.org/ pub_releases/2012–08/tkai-akr080612.php, 2012.

Zhao, Y., A. Meek Garrett, G. Levine Benjamin, R. Lunt Richard, Near-infrared harvesting transparent luminescent solar concentrators, *Advanced Optical Materials*, 2014, 2 (7), 606–611.

Design Case 4

Design of a PV-Powered Racing Car

Merel Oldenburg, Michael ten Den, and Angèle Reinders

4.1 INTRODUCTION TO THE SOLAR TEAM TWENTE AND THE BRIDGESTONE WORLD SOLAR CHALLENGE

Every two years a brand-new solar car is developed by Solar Team Twente to participate in the Bridgestone World Solar Challenge and hence contribute to a futureproof world by the development of sustainable transportation (Figure DC-4.1). In 2019, the eighth generation of Solar Team Twente consisted of 19 students from the University of Twente and Saxion University of Applied Sciences, both located in Enschede, The Netherlands. Together with its large network of innovative partners from business, it created a space that fostered innovative thinking, creation, and testing. To maximize the use of this space, the team has given itself a grand challenge: to win the Bridgestone World Solar Challenge.

Until 2019, the Solar Team Twente has developed seven cars so far, the eighth was revealed in June 2019 (see Figure DC-4.1). A solar car is powered by a battery-powered electric engine that is charged by a solar panel integrated in the car (Figure DC-4.2). The solar panel is made of interconnected solar PV cells. In its relative short history, the team has been successful on multiple occasions. This includes being two-time European champion as well as belonging to the top teams competing in the World Solar Challenge.

FIGURE DC-4.1 RED-E: The 2019 solar racing car of Solar Team Twente.

FIGURE DC-4.2 How many solar cars can you find on this photo? This photo from 2015 shows the preparations for an exhibit of the five existing solar cars, which are all partially visible.

4.2 THE BRIDGESTONE WORLD SOLAR CHALLENGE[1]

Since the first edition in 1987, the international Bridgestone World Solar Challenge (BWSC), which takes place every two years in Australia, has demonstrated the possibilities of developing alternative automotive technology. The BWSC challenges young people from all over the world to join this race that travels from the North to the South of Australia, starting in Darwin and finishing 3,000 km later in Adelaide (Figure DC-4.3). During the race, cars may solely be powered by the Sun.

The BWSC describes its purpose[2] as: "stimulate research into, and development of, sustainable road transport. The World Solar Challenge is primarily a design competition. The regulatory philosophy is to provide the parameters on which to base the design, rather than specify exactly how to build a solar car. Science and technology evolve and to encourage the most innovative ideas, event requirements also evolve." The regulations form the basis on which cars are designed; hence, they form the main design challenge for the teams.

Some major developments can be seen when looking at the regulations for different years. The evolution of the regulations can be clearly observed in the design of the three Solar Team Twente cars shown in Figures DC-4.4 through DC-4.6, indicating a decrease in solar array surface and an increase of user friendliness.

---

[1] https://www.worldsolarchallenge.org/about_wsc/history.

[2] 2019 regulations.

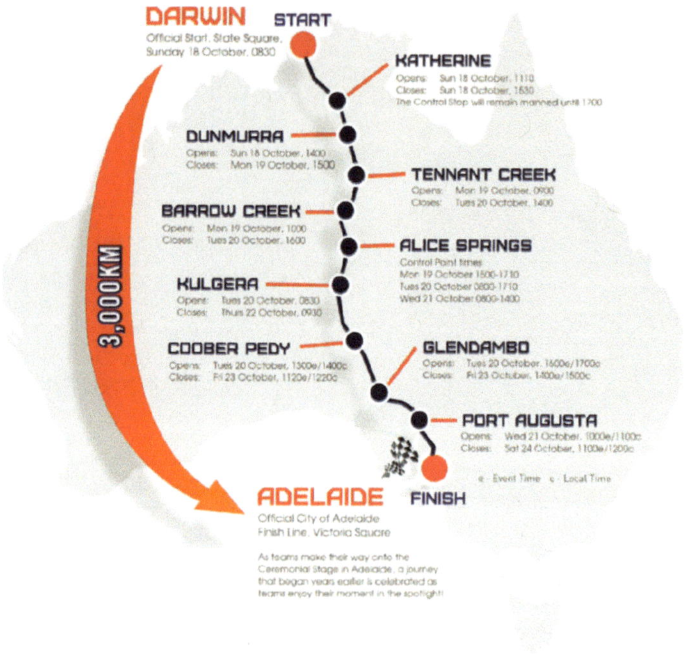

FIGURE DC-4.3 The route map of the 2015 race shows that the World Solar Challenge stretches over 3,000 km, from Darwin to Adelaide. Along the way, teams have to stop nine times for a "control stop" as indicated on this map.

FIGURE DC-4.4 The first solar car of the Solar Team Twente was released in 2005. It had a 9 m^2 GaAs PV array, three wheels, and to step into the car the driver needed support by at least six people. Please notice that 2019 regulations prescribe a maximum of only 2.6 m^2 GaAs solar PV cells.

FIGURE DC-4.5 In 2013, the Solar Team Twente participated in the BWSC with the RED Engine, which has a 6 m^2 silicon PV array, and it takes "just" four people to close the PV panel. Please notice that this was the first solar car of the Solar Team Twente with four wheels.

FIGURE DC-4.6 RED Shift, the 2017 car, was a lot smaller than the previous cars, having a 4 m^2 silicon solar panel. This was also the first solar racing car in which the driver was able to enter and close the panel without any help of team members.

In 2019, the BWSC had three classes in which teams may participate: the Challenger, Cruiser, or Adventure class. Solar Team Twente has been participating in the Challenger class, which concerns single-passenger race cars. It focuses on cutting-edge technology and innovation to create faster cars, for which reason it can be seen as the Formula 1 of solar car racing. The relatively new Cruiser class is meant to bring forward efficient and practical cars that can seat at least two drivers. Cars are designed to be road legal and should speed up the gap between conventional and solar cars. The Adventure class is meant for solar cars that were actually designed for previous events but which were not eligible the BWSC in the Challenger or Cruiser class within the existing regulations.

4.3 CHALLENGES OF BUILDING A SOLAR CAR

Apart from the World Solar Challenge itself, there exist a more general global challenge when developing solar cars. The Solar Team Twente, together with the whole world solar team collective, hopes to soon see solar cars in daily traffic. But before that will happen, solar cars must become a comparable alternative to current conventional cars on the market. This prerequisite is not related to design, comfort, or readiness for mass production but merely based on core functional features of cars such as their power, speed, and drive range.

At the end of the nineteenth century, the first cars with internal combustion engines were designed. For the first models, the maximum speed was just 30 km/h, while nowadays they can sometimes reach 400 km/h. However, the engineering principles behind the force balance of a vehicle have remained the same:

$$F_{\text{generated}} = F_{\text{resistance}} + F_{\text{drive}}$$

where the force generated, $F_{\text{generated}}$, by the engine and power train, is used to overcome resistance forces, $F_{\text{resistance}}$, and for propulsion forward by driving force, F_{drive}. The resistance has multiple components, from which drag is the most dominant one. Hence, the chassis of a car, and also *that of* a solar car, directly influences the resistance and therefore multiple features, such as the car's speed and drive range. In the early days, the relations between speed and design were essential regarding the development of faster cars. Since then, the car market has grown exponentially and engine design has significantly improved. Because of the existence of high-power engines, the relation between speed and the design has become negligible for conventional cars. In other words: nowadays it barely matters if the car (chassis) design is not efficient since it can be easily overpowered by its internal combustion engine (ICE).

Unfortunately, for solar cars the relation between speed and design is still valid. However, this gap compared to conventional passenger cars with an ICE is quickly closing and the first solar cars are already legally admitted to public roads. In essence, a solar car is an electrical vehicle (EV) with an internal charging system, namely the PV system that is integrated in the car's body parts. The main disadvantages of electric cars are a relatively slow recharging process by electricity from the grid and a limited battery capacity resulting in a limited drive range of 250–400 km. A solar car tries to mitigate these issues by adding another source of energy, besides the grid, namely the Sun. This causes new challenges because of the intermittency of solar irradiance, which changes with the moment of the day and the weather. Moreover, the Sun isn't shining at night. This requires special design features of solar-powered EVs. In this design case, it is therefore discussed how Solar Team Twente dealt with the challenge of finding the balance between power consumption by the car and power production using solar irradiance in the design and further development of their racing car. As an example, the best car so far, the RED One, will be explored and explained.

4.4 THE FRAMEWORK OF THE DESIGN OF A SOLAR RACING CAR

In October 2015, the RED One was put to its ultimate test, the 2015 World Solar Challenge, in which it finished only a few minutes after its victor. The car showed its strength again by winning the European Solar Challenge in 2016 and becoming second in 2018 (just behind RED Shift, Solar Team Twente's 2017 car). This makes the RED One Solar Teams Twente best performing solar car so far.

Only 13 months before the 2015 BWSC, in September 2014, the team started with their design process. While nothing was designed yet, there was already a strong framework in place from the previous teams.

The basis of the design framework is the 2015 BWSC regulations,[3] which were published in June 2014. Compared to the 2013 regulations, only a few changes had been applied to the 2015 regulations. Meaning that, in general, a lot of room existed to further improve the new car's design on the basis of the 2013 car design. Throughout this case, it is shown how this happened.

The most important regulations for the 2015 BWSC form the basis of the design of solar cars. These are:

- Rule 2.1.2
  Challenger Solar EVs are designed for efficiency. They carry a driver only. The winner of the Challenger Class will be the first Challenger Class Solar EV to complete the course in accordance with the regulations.
- Rule 2.2.1
  Challenger Class and Cruiser Class Solar EVs must be no more than 4,500 mm in length, no more than 1,800 mm in width, and no more than 2,200 mm in height (above the ground) at any time while charging or driving.

[3] BWSC regulations version 1.2, 12 February 2015.

- Rule 2.3.1
  Challenger Class and Cruiser Class EVs must be supported by four wheels: two front wheels and two rear wheels.
- Rule 2.8.1
  Teams must demonstrate that all occupants (one per seat) can exit the solar EV in less than 15 seconds without assistance.
- Rule 2.21.6
  For Challenger Class and Cruiser Class Solar ECs, if the collar collector uses photovoltaic cells, then the allowable area of photovoltaic cells is:
  - Not more than 6.000 square meters for solar EVs using only silicon photovoltaic cells
  - Not more than 3.000 square meters for solar EVs using only GaAs photovoltaic cells

  For Challenger Class and Adventure Class Solar EVs, if the energy storage system is a secondary electrochemical battery, then the sum of the nominal cell masses, as specified and endorsed by the cell manufacturer and approved by the Chief Energy Scientist, may not exceed the following limits:
  - Li-ion 20.000 kg
  - Li-Polymer 20.000 kg
  - LiFePO4 40.000 kg
  - Ni-MH 70.000 kg
  - Pb-Acid 125.000 kg.

4.5 PROJECT PLANNING

An important element in the design framework of a solar car is technical project planning. Ideally, several design cycles have been made before starting the production of a car's design. However, due to the limited timespan, there is no time for endless design cycles. In general, the following rule applies: "If you fail to plan, you plan to fail." Therefore, in order to achieve the best result in the limited time available, an iterative V-model is used. This model allows the solar team to relatively quickly test design concepts regarding their feasibility, which significantly improves the development speed and reliability of the solar car.

The technical planning consists of five phases:

1. Concept phase
2. Design phase
3. Mock-up phase
4. Production phase
5. Test phase

And of course a sixth phase: the race itself.

Most relevant for the design of the car are the first three phases. In the first phase, the concept phase, several designs are investigated for each domain (electrical, aerodynamic, etc.). By developing models and using previous race data, a profit-risk ratio can be determined for each design, and the next-best conceptual design can be selected for further development. The risk assessment is compulsory to minimize the risk of failure of new, previously never-developed designs that have a great potential to improve the car. New innovative ideas are essential to be able to win the race; however, they should not affect the reliability of the car. These new ideas can be related to the choice of the type of PV cell choice, as well as the aerodynamic concept, which are both interconnected if it comes to the final performance of the car.

At the end of the concept phase, all major design concepts have been selected.

In the next phase, the design phase, all concepts will be further developed. In the design phase, simulations with many iterations are executed until a complete car design is realized. For the aerodynamic design, this is the final design, whilst the designs of the mechanical and electrical components will be enduring the mock-up phase.

In the mock-up phase, an aluminum mock-up car is used to test the mechanical and electrical components in practice (Figure DC-4.7). This is necessary because on the basis of the simulations it can't be guaranteed that all components interconnect and fit as intended. Therefore, in the mock-up phase, adjustments in the mechanical and electrical design are made accordingly. In 2015, this phase also included a test in the wind tunnel with a scale model to evaluate the aerodynamics of the car—first to discover flaws that could be repaired during the production phase and second to serve as verification of the aerodynamic models used by the team (Figure DC-4.8).

FIGURE DC-4.7 A drone shot from February 2015 showing the 2015 mock-up car, which was tested regarding its mechanical and electrical components at airport Twente.

FIGURE DC-4.8 A scale model of the RED One was tested in a wind tunnel. The models used to design the car were compared to the airflow simulated in the tunnel.

4.6 ENERGY MANAGEMENT

During the design process of a solar car, a strong focus is put on the energy management of the car. Energy management optimizes the balance between energy generation and consumption. The regulations of the BWSC state that during the race the only source of energy is solar energy. All solar cars therefore start with a fully charged battery pack; however, they're not allowed to charge the battery pack during the race using other sources of energy than the Sun.

The challenge of the team is to design a solar car wherein energy consumption is minimized and energy generation is maximized.

The general equation for energy input, E_{in}, and energy consumption, E_{out}, simply is as following:

$$E_{out} = E_{in}$$

In the ideal situation, with no wind, drag, gravity, and 100% efficient electronics, all incoming energy (E_{solar}) would be directly transferred to the motor. This will substitute the input energy and the equation becomes:

$$E_{out} = E_{solar}$$

The motor consumers energy, E_{motor}, so substituting this for the output energy gives:

$$E_{motor} = E_{solar}$$

However, in reality, the situation is not ideal; therefore, some extra terms should be added to the equation, to start with the rolling resistance. The wheels of the solar car are pressed on the road by gravity, which causes deformation of the wheels, which causes a resistance. This costs energy, E_{RR}; therefore, it is part of the following energy balance formula.

$$E_{motor} + E_{RR} = E_{solar}$$

This already shows that not all the energy of the solar panel is going to the motor, but also a bit to the rolling resistance. By minimizing the rolling resistance, the available energy for the electric motor is maximized. This results in faster propulsion on the road. The same goes for other resistances.

When driving, air particles collide with the surface of the car and slow down the car. This is called air drag, E_{aero}, and the equation becomes:

$$E_{motor} + E_{RR} + E_{aero} = E_{solar}$$

The bigger the surface, the more air particles have to be moved, hence the greater the drag. By making the car more aerodynamic, more streamlined, less air particles have to move, and the particles that do can be better guided around the car. This will result in less energy consumption.

Finally, there is another type of loss that should be included in the equation: electrical losses. In order to transform the energy from sunlight into motor power, a lot of power electronics is required. Solar cars are subjected to the efficiency and energy losses of electrical components, $E_{electrical\ loss}$, so the full energy equation thus becomes:

$$E_{motor} + E_{RR} + E_{aero} + E_{electrical\ loss} = E_{solar}$$

In order to be able to move forward, the power train of the car needs energy to be transformed into movement. In the case of a solar car, two sources can provide this energy: the solar panel and battery pack.

4.6.1 The Solar Panel

Energy from the Sun is captured by the solar panels. The Sun is theoretically a sufficient and never-ending source of energy for the solar car, namely when only considering incoming and outgoing energy, RED One should be able to keep driving forever when driving in perfect conditions at a cruise speed of about 90 km/h.

All incoming energy from the solar panels is routed toward the power train of the car. If this energy exceeds the energy that the powertrain needs for drive, this excess energy can be stored in the battery pack (Figure DC-4.9).

4.6.2 Battery Pack

The battery pack forms another energy source that is required, since actually driving in perfect conditions will never occur in reality. When there is too

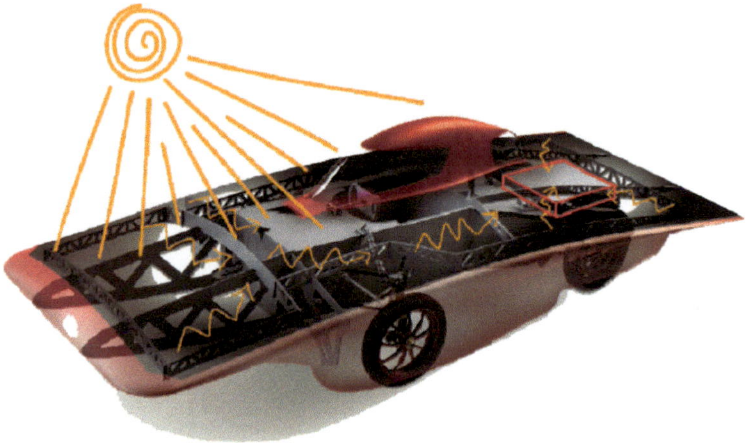

FIGURE DC-4.9 A schematic overview of solar energy captured by RED One during static charging. Solar energy is captured all over the panel and routed toward the battery pack (in red).

FIGURE DC-4.10 A schematic overview of the energy flow from the battery to the motor.

much energy coming in from the Sun, the excess energy is saved in the battery. When solar energy is insufficiently available, additional energy from the battery will be used to power the car. In the case of static charging, all energy captured from the solar panels will be directly used for charging the battery (Figure DC-4.10).

Let's look at an example to explain whether a battery pack is actually required to be able to drive a solar car. To find an answer, one can compare the driving speeds in two totally different situations: driving on a sunny day in Australia and driving on a winter day in the Netherlands. In this example, it is assumed that electrical losses don't exist and that there is no battery pack.

We will first explore the resistance forces out of which the largest one is drag, F_{drag}, represented by this simple model:

$$F_{\text{drag}} = 0.5 * C_d A * \rho * v^2$$

Second, rolling resistance can be represented by:

$$F_{\text{rr}} = C_r * F_N = C_r * m * g$$

The total resistance force consists of drag and rolling resistance, in the following way:

$$F_{\text{resitance}} = C_r * m * g + 0.5 * C_d A * \rho * v^2$$

The energy input only consists of the energy generated by the solar panel, which can be determined using the irradiance (W/m^2), PV efficiency (%), and surface of the solar panel (m^2). In this case, we use the relation between power, *P*, force, *F*, and speed, *v*, which is given by:

$$P = F * v$$

When comparing RED One, driving in Australia or the Netherlands, the following numbers can be used:

On a summer day in Australia, irradiation can reach 1000 W/m^2, and in the Netherlands a winter day might give you 250 W/m^2 irradiation. RED One has a solar panel of 6 m^2 with an efficiency of 24%. Including a driver, RED One weights 230 kg; for the C_r, 0.006 can be taken and 0.13 for the $C_d A$. Could you figure out the speed of the cars in both situations? What does this say about the importance of a battery pack?

4.7 DESIGN SOLUTIONS OF THE RED ONE

The regulations from the World Solar Challenge always forces its participating teams to face the energy balance dilemma while designing the car. In 2015, the allowed surface of PV arrays decreased, as well as, the capacity of the battery pack. Therefore, in the design process of Solar Team Twente, two design results were aimed at: maximizing energy production and minimizing energy demand. The following three examples show how Solar Team Twente achieved these results by design solutions of the RED One.

4.7.1 Aerodynamics

The largest loss in a solar car is the drag. To optimize drag, during the design and detailing phase (see project framework) three engineers have been working full-time on the optimizing the aerodynamic design of the 2013 car (Figure DC-4.11). Diving deep into the aerodynamics would lead to a whole book of its own.

FIGURE DC-4.11 The RED Engine (2013) clearly has a curved PV panel.

FIGURE DC-4.12 The solar panel of the RED One (2015) has barely any curve left to decrease drag.

Therefore, without going into theory, one big visible improvement was applied when designing RED One (see Figure DC-4.12), namely the use of a non-curved solar panel in the 2015 car. A curved panel and a non-curved panel hardly yield a different energy performance; however, in practice, for the placement and performance of the solar cells, a non-curved panel turned out to be much more practical. Hence, in 2015 there is almost no curve left in the panel while at the same time drag has been minimized.

4.7.2 Body Design

Most of the time, a new design concept that maximizes solar energy productions tends to create energy losses elsewhere in a solar car, and vice versa. But sometimes an idea comes along that both maximizes production and minimizes losses; in 2015, this was the case with the upper body design of the car. It carries the most important feature of a solar car—the solar panel. For the RED One's upper body, it was decided to add cavities underneath the solar panel to reduce temperature losses in the solar cells, which simultaneously resulted in weight reduction of the car, leading to less energy losses during driving (Figure DC-4.13).

4.7.3 Asymmetric Concept and Solar Cell Selection

One type of constraint faced when designing a car for the BWSC is the prescribed route on the roads between the start and the finish. This in combination with the course of the Sun has to be taken into account when optimizing a solar car design. A major design choice in the RED One can be led back to these specific condition during the race.

Namely, one of the biggest issues with the design of an optimal solar panel is the occurrence of shades originating from the canopy and other objects. The Solar Team Twente cars must have a canopy in order to comply with the regulations. In 2015, when looking at the course of the Sun in combination with the

FIGURE DC-4.13 The bottom of the upper body of RED One with integrated solar cells. When looking though the cavities, the solar cells can be seen through the support structure.

FIGURE DC-4.14 The two versions of solar cells on the RED One: cut and uncut cells.

route, a major design choice was made to place the canopy at the right side of the car and as such minimize the shaded area on the solar panel; an extra feature that was added to the RED One was the use of both cut and uncut cells. When part of a cell is shaded, the whole cell's performance is affected. Hence, having more cells in the same area minimizes the effect of shading. Therefore, in the area around the canopy, cells were cut *into* smaller cells as shown in Figure DC-4.14.

4.8 CONCLUDING WORDS

Designing a solar car is such a comprehensive process that this design case shows only the tip of the iceberg of all design aspects and design decisions. It however served as an introduction to the Solar Team Twente project, the design framework, and one of the main challenges: balancing energy production and energy consumption in a solar racing car. A challenge that was once faced too by designers in the nineteenth century when developing the first internal combustion engine cars. The Bridgestone World Solar Challenge possess a challenge that pushes teams from all over the world to the limit and toward innovative solutions in the field of solar-powered EVs. Solar Team Twente believes that if all teams continue to develop technology in this fashion, it is only a matter of time until you will notice a solar car pulling up next to you on the highway, and maybe it will be the Lightyear or the Sion (see Figures DC-4.15 and DC-4.16).

FIGURE DC-4.15 The Lightyear One, a PV-powered EV, which is commercially produced by Lightyear in the Netherlands. (Copyright by Lightyear.)

FIGURE DC-4.16 The Sion, a PV-powered EV, which is commercially produced by Sono Motors in Germany. (Copyright by Sono Motors.)

5 Building-Integrated Photovoltaics

Eelke Bontekoe, Wilfried van Sark, and Joost van Leeuwen

CONTENTS

5.1 INTRODUCTION

Photovoltaic (PV) technology has shown a turbulent trajectory in research, development, and deployment, which has led to an expected 0.5 terawatt-peak-installed capacity globally at the end of 2018 (IEA PVPS 2019). Today, prices have more or less stabilized, and market volumes show a healthy growth while national support schemes are being reduced or redefined, which all exemplify that PV is well on its way to becoming a major player in the supply of renewable energy in the coming decades (Breyer et al. 2017). Most of the installed capacity in the built environment can be found on roofs of buildings in residential and commercial districts. In residential areas, PV is usually attached or added to the roof, and therefore one generally refers to this as building-attached PV (BAPV). Integration of PV in the building envelope (roof and/or façade) is then referred to as building-integrated

(a) (b)

FIGURE 5.1 (a) BIPV: PV roof tiles integrated into the roof and hardly visible. When PV modules are removed, new tiles must be placed to maintain water tightness of roof. (Solinso 2018b); and (b) BAPV: PV modules are placed on top of the roof tiles. PV modules can be removed without losing other roof functions than electricity generation. (©W. van Sark)

photovoltaics (BIPV). This chapter provides many examples of BIPV; in Figure 5.1 the difference between BIPV and BAPV is illustrated for a roof.

BIPV has been emerging as an interesting market segment (Jelle et al. 2012; Frontini et al. 2015; Ritzen et al. 2017; Zanetti et al. 2017; Osseweijer et al. 2018). The size of the global BIPV market was about 2.3 GW (or ~1% of the global PV market) (Global Industry Analysts 2015) in 2015, with Europe constituting the largest market (42% of global market) in particular due to attractive incentives in France, Italy, and Germany. On a global level, Europe and the United States dominate the BIPV market, while Asia is addressing BIPV as well (Shukla et al. 2017). A recent market report showed that the present BIPV market size amounts to $2.9 billion in 2018, of which China accounts for about one-third (n-tech Research 2018). It is projected that the market size in the US will increase from about $0.8 billion in 2018 to $4.4 billion in 2027. The EU market is expected to grow from $0.25 billion to $2.5 billion EU, while in China the market size will remain nearly constant. The global market size in 2027 is estimated to be about $11.6 billion. To date, the BIPV market is still a niche market. However, based on the European Energy Performance of Buildings Directive (EPBD) 2010/31/EU (European Commission 2010), which is being renewed in the so-called "Winter Package" (European Commission 2016), a tenfold increase of the European BIPV market is estimated for 2020, which is even faster than predicted by n-tech (n-tech Research 2018). Note that the present directive states that all new buildings of the 28 EU member states should be nearly zero-energy buildings (NZEBs) by 2020. The EPBD states that "the nearly zero or very low amount of energy required by the building should be covered to a very significant extent by energy from renewable sources, including energy from renewable sources produced on-site or nearby." Clearly, one possible solution to realize NZEBs is the generation of renewable electricity on-site, by means of BIPV. Another driver for BIPV is the attractiveness of BIPV in terms of aesthetics and flexibility, which may contribute to market growth too (Osseweijer et al. 2018). Several barriers need to be surpassed though, which relate to integration of the PV industry and the building industry (Krawietz et al. 2016). This may hamper the relative market size of BIPV to the market size of PV itself. Interestingly, about two-thirds of the European BIPV market is realized in new buildings, and one-third in renovation (Delponte et al. 2015). Façade BIPV accounts for half of the projects. A recent inventory shows hundreds of BIPV products offered (Zanetti et al. 2017).

One main challenge in the field of BIPV is the combination of electricity generation with design requirements for building components to be used in building envelopes. To alleviate this challenge, standardization for BIPV systems is developed in EN 50583 (CENELEC 2016a, 2016b). Besides electricity production, BIPV fulfils similar functions as structural building elements, such as sound protection, thermal control, and weather proofing. Additional functions such as shading, aesthetics, and strength are also realized. On top of that, generating decentralized energy within the built environment aids to provide a redundancy of the current electricity network, defines a state of independence, improves energy efficiency in the building, and avoids transportation losses in electricity grids (Scognamiglio 2017).

Adding PV on the roofs of buildings, especially when these buildings are taller than about three stories, may not be generating sufficient energy to meet the full building electricity demand. Façades offer additional potential for PV (Defaix et al. 2012), either applied to the façade surface itself (building added/attached/applied PV, BAPV), or integrated in the façade (BIPV). In particular the use of large glass façades is a growing trend. Market predictions show that glass-applications (façades) will develop faster than roof-based applications (Ballif et al. 2018). Specific BIPV designs and engineering have been reviewed recently see, for example, Heinstein et al. (2013), Ferrara et al. (2012, 2017), Tripathy et al. (2016), Biyik et al. (2017), Efurosibina Attoye et al. (2017), Jelle et al. (2012), Ritzen et al. (2017), Zanetti et al. (2017) and references therein. BIPV applications offer the possibility to replace regular building envelope components into prefab-integrated components that at the same time generate electricity (Ritzen et al. 2017). BIPV applications are flexible in size, shape, color, and appearance and can be combined with materials commonly used in construction, such as glass or metal (Frontini et al. 2015). This contributes to the aesthetic value of a building, allowing flexibility in architectural design (Farkas et al. 2015). Performance of BIPV in urban areas depends on orientation and tilt, just like PV panels on roofs, but more importantly on the surrounding buildings or other shadow-casting structures. Modeling of performance to predict annual yield of BIPV is more complex than regular PV (Lee et al. 2014).

In the following, this chapter discusses technical aspects of building envelopes, shows examples of BIPV components (modules, balance of system and mounting systems) and connects these aspects to their energy performance.

5.2 TECHNICAL ASPECTS OF BIPV BUILDING ENVELOPES

BIPV is defined as being essential elements in the building envelope that contain at least one additional function besides generating electricity (e.g., insulation or exterior weather barrier). Without the BIPV element, a non-BIPV building material or component would be required to replace it. Thus, power generation by a BIPV component can be considered secondary to the role of being a building material or structural component, while it is essential in generating locally energy that is demanded by the building. The BIPV system is limited to the external structure of the building in order to receive sufficient solar energy. Apart from structural integration, it has little limitations, mostly related to maximize harvesting the incident solar irradiance. For this reason, façades and roofs are both ideal locations for BIPV, and there are different types of systems available for both façades and roof. In designing a BIPV system, one needs to consider: (1) optimal conditions for energy production, (2) architectural design (aesthetics), (3) requirements of construction materials (insulation, ventilation, etc.), (4) financial aspects, and (5) buildings regulations.

Building requirements have evolved over the past centuries. A recent review (Clua Longas et al. 2017) has analyzed the evolution of façade systems through history, which in general also holds for roofing systems, in an attempt to define near-future façade requirements. Figure 5.2 illustrates that protection from the environment was the first and foremost function of a façade and roof since the industrial revolution, while after the oil crises and the present energy strategies toward greenhouse-gas

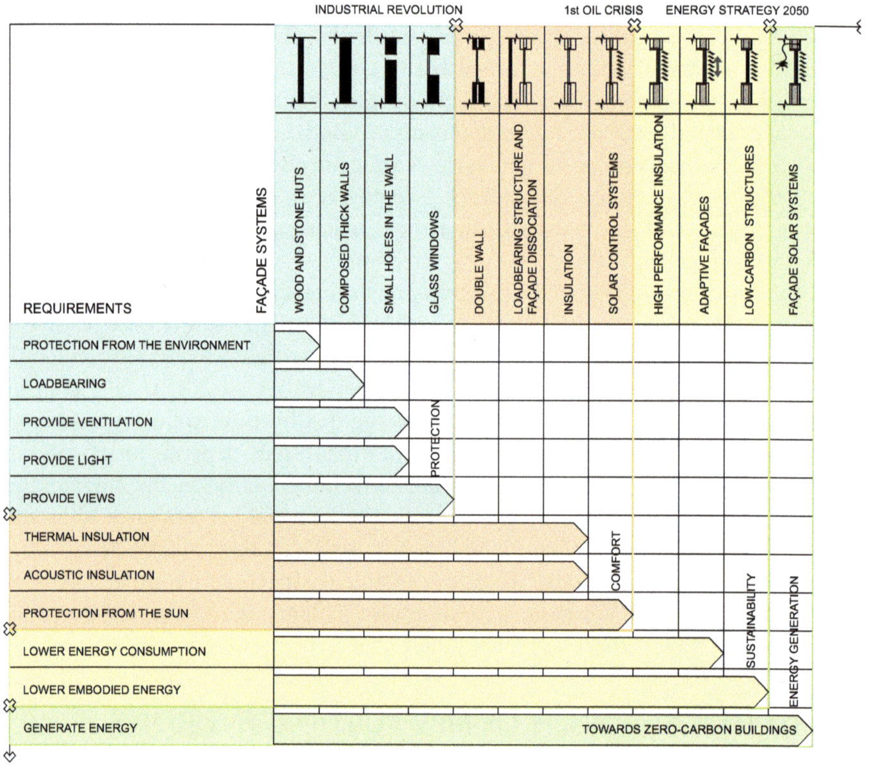

FIGURE 5.2 Residential façade requirements in historical context. (From Clua Longas, A. et al., Towards advanced active façades: Analysis of façade requirements and development of an innovative construction system, In *PLEA 2017, Edinburgh, Design to Thrivem*, 2017.)

emission reductions by 2050, generating energy has been added. Thus, achieving a low- or zero-carbon building adds the following requirements: (1) minimizing the building's energy consumption, (2) lowering its embodied energy, and (3) generating energy as part of the envelope solution (Clua Longas et al. 2017).

5.3 BIPV APPLICATION EXAMPLES

BIPV applications generally can be classified as either being applied in façades or roofs. A further subclassification can be made, as shown in Figure 5.3 (Frontini et al. 2015). Regarding façade systems, we distinguish in general between cold and warm façades (or curtain walls), while regarding roofs, a distinction is made between pitched roofs (at certain tilt) and flat or curved roofs. Table 5.1 gives a short comparative analysis between the main classifications of BIPV products.

5.3.1 Façade Systems

A **cold-façade** means that the BIPV-system is not directly responsible for the indoor-temperature of the building. For example, an insulation layer or ventilated area is present behind the PV modules. Figure 5.4 shows a very early example: the CIS tower in Manchester, UK, which is the first European large-scale example of BIPV. In 2006, this building was retrofitted with over 7,000 crystalline silicon solar panels, generating 180 MWh/year (Roberts and Guariento 2009).

Figure 5.5 shows a sketch of a "rainscreen overcladding" system. The outer leaf is the major barrier to rain penetration. Behind these rainscreen panels, and an air cavity, the inner leaf of the system acts as a barrier against the wind. Besides protection against heavy wetting, the rainscreens also block solar radiation on the inner leaf and serve as a cosmetic element.

A BIPV system is realized when a PV module is integrated into the rainscreen panel, as shown in Figure 5.6. The ventilated cavity behind the panels aids to limit the operating temperatures of the BIPV components; further on, a connection box is added. In case of existing cladding technology, for example, during renovation, no major modification on the building is required to incorporate solar modules. Figures 5.5 and 5.6 also illustrate that an extra insulation layer outside of the backing wall serves as the main heat barrier, hence the name, cold façade.

Opposite to the early example of Figure 5.4, where crystalline silicon PV modules are clearly visible, new developments in BIPV technology lead to more design freedom for architects. An example is the Kuijpers building in Helmond (Panovenweg 20, Netherlands). The façade was designed by Studio Solarix (2019) and manufactured and installed by Sorba (2019), with Kuijpers (2019) as technical partner (Figure 5.7). The folded aluminum/composite façade elements are equipped with monocrystalline silicon PV cells. There are three different elements with different colors and sizes. These glass/glass solar modules are supplied by Kameleon Solar, and the color is printed on top of the glass with their colorblast technology. This technology allows one to print different colors and patterns on solar modules while maintaining a high light transmittance on the solar cells. The cells are invisible from a distance of more than 5 m (Kameleonsolar 2019) and generate about 8,000 kWh/yr.

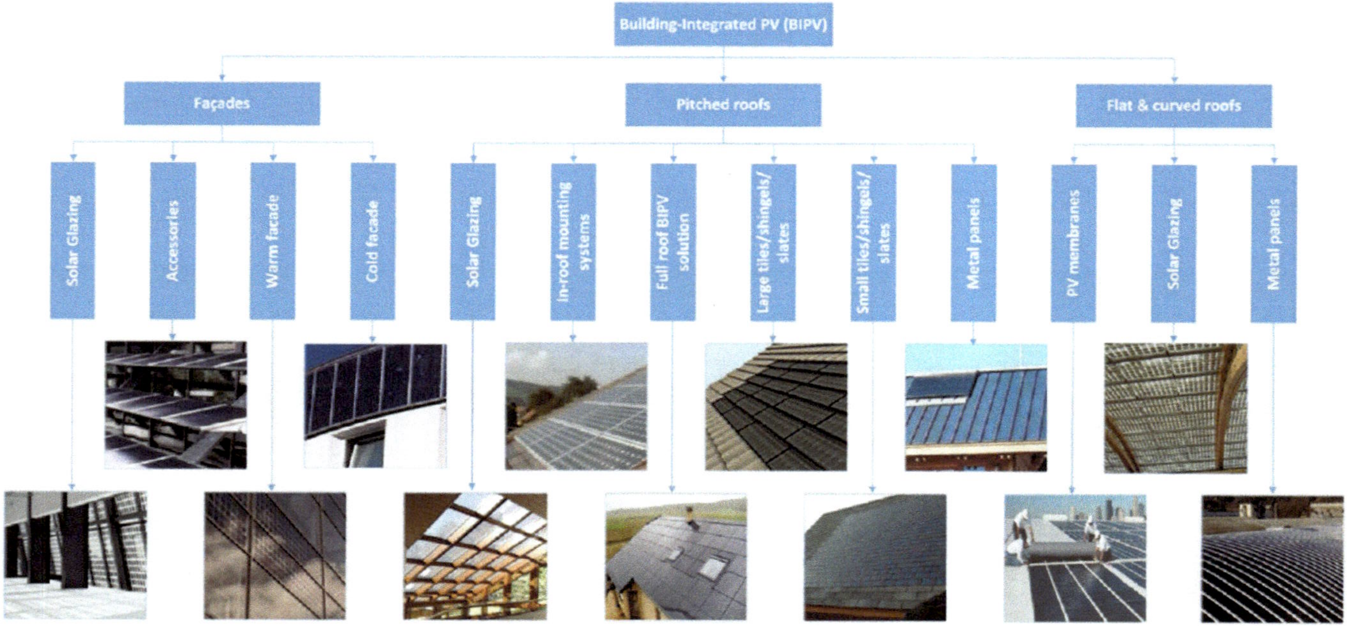

FIGURE 5.3 BIPV classification into façades and roofs. (From Frontini, F. et al., BIPV product overview for solar façades and roofs—BIPV status report, 2015.)

TABLE 5.1
Comparative Analysis between Different BIPV Products

Product	Advantages	Disadvantages	Applications
Standard in-roof systems	i. Easy to handle ii. Competitive iii. Higher efficiency/ Performance	i. Less esthetic ii. Used only for certain roof types iii. Multifunctional aspects of PV are absent	Pitched roofs of residential and commercial buildings
Semi-transparent system	i. Most unobtrusive and most esthetic BIPV. ii. Used in prestigious buildings with well-visible façades and skylights. iii. Marginal daylight elimination/capacity to diversify light intake. iv. BIPV cell shapes can be attractive. v. Thin-film BIPV cells have uniform appearance and suitable for flush mounting	i. The units are very heavy ii. The prices are normally high since they are usually tailor-made products iii. Seamlessly integrated and difficult to notice the presence of PV modules iv. Difficult for hiding the cables v. Sizes and shapes of cells are limited vi. Silver tabbing interferes the transparent spaces	Semitransparent façades, skylights, and shading systems in commercial and public buildings
Cladding systems	i. Suited if the PV system is to be recognized ii. Different colors and visual effects can be incorporated iii. More efficient	i. Low system performance ii. The lower parts of façades are normally not used due to possible shadows iii. Installation cost is higher	External building walls and curtain walls in commercial and public buildings
Solar tiles and shingles	i. Esthetic for residential pitched roofs ii. Higher efficiency iii. Light and easy to install	i. Small unit sizes. Hence, longer installation time ii. High cost-performance ratio iii. Prone to break	Pitched roofs in residential buildings and old buildings
Flexible laminates	i. Lightweight (most suitable for weak roofs) ii. Easy to handle and install iii. Low balance of system cost iv. Used on roof surface v. Curved installation possible	i. It doesn't fulfill other functions of building components. ii. Efficiency is low	Flat and curved roofs of commercial and industrial buildings with large unused roofs

Source: Tripathy, M. et al., *Renew. Sust. Energy Rev.*, 61, 451–465, 2016.

FIGURE 5.4 View from the south onto the CIS tower in Manchester, UK. (Courtesy of Stephen Richards/Co-operative Insurance Society Tower, Miller Street, Manchester (2)/CC BY-SA 2.0.)

Another example of a new development in BIPV façade elements is shown in Figure 5.8. The picture shows a closeup and a test setup of 3D-shaped fiber-reinforced façade elements. Copper indium gallium diselenide (CIGS) thin-film cells are laminated into the material.

In a **warm façade**, also referred to as "curtain wall," the outer-structure of the building envelope is directly connected with the inside of the building space. Therefore, it is actively responsible for insulation as well. To minimize the influence of the difference between the exterior and the indoor temperature, a low heat transmittance of the warm façade is required. Both the vision area as the spandrel area of the façade can be utilized to install PV modules. Installation for PV is comparable as for ordinary glass. Single or double-glazed units can be replaced by (semi-)

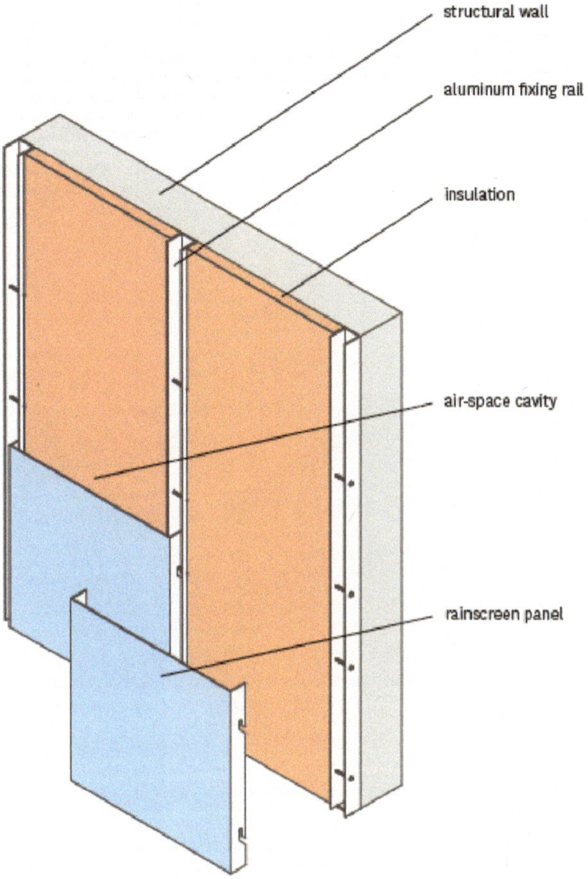

FIGURE 5.5 General scheme of a rainscreen cladding system in front of an opaque wall. (From Roberts, S., and Guariento, N., *Building Integrated Photovoltaics: A Handbook*, Walter de Gruyter GmbH, 2009.)

transparent or opaque, single- or double-glazed PV modules. Figure 5.9 shows the detail of how a unitized curtain wall system with double-glazed PV modules could be constructed.

Different PV technologies are available for these (semi-)transparent PV glazing units. Figure 5.10a shows an example of PV glazing with crystalline cells. The PV cells in this example are the opaque squares in the glass. In this case, the size of the cells and the spacing between the cells define the transparency of the module. When

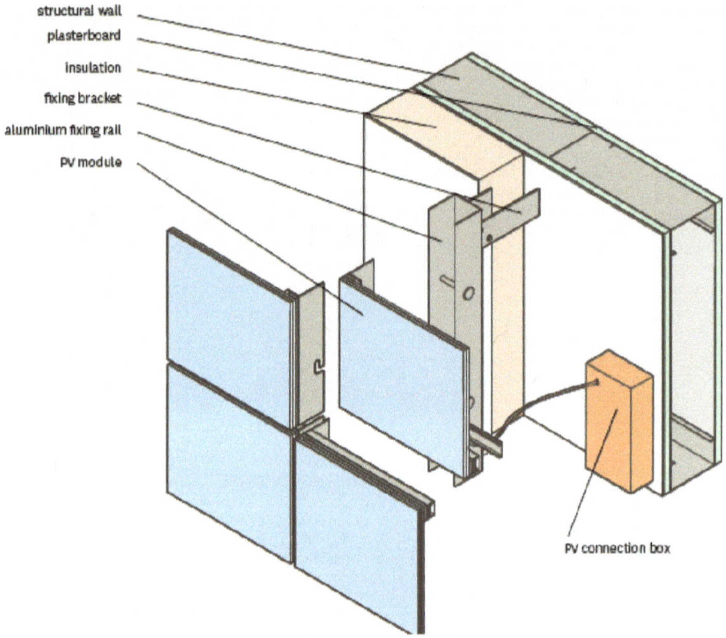

FIGURE 5.6 Detail of a rainscreen panel integrated with a PV module. (From Roberts, S., and Guariento, N., *Building Integrated Photovoltaics: A Handbook*, Walter de Gruyter GmbH, 2009.)

FIGURE 5.7 Kuijpers building Helmond (The Netherlands), the south façade is equipped with 177 custom-made BIPV elements. A consortium consisting of Kuijpers, Studio Solarix, Sorba, and Kameleon Solar was responsible for the design and realization of this building.

(a) (b)

FIGURE 5.8 (a) Flexipol 3D shaped fiber reinforced façade element with thin-film solar cells (CIGS) laminated into the material; and (b) test setup with Flexipol façade elements. (From Flexipol, https://www.flexipol.nl/workshop-voor-architecten-bipv/, 2019.)

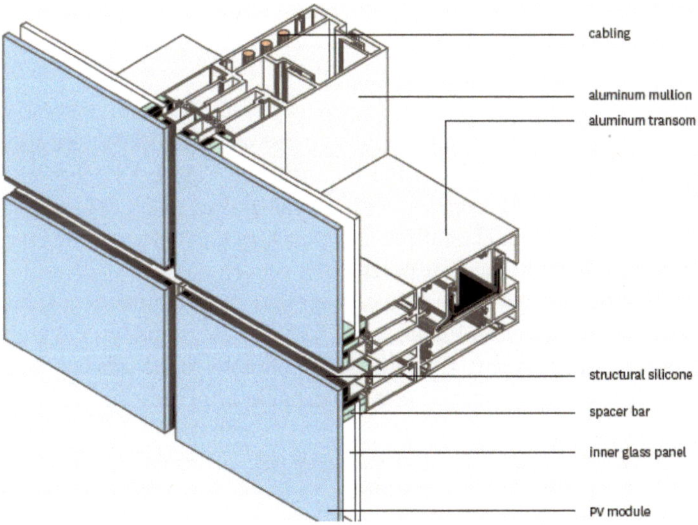

FIGURE 5.9 Detail of a warm façade system with PV modules in double-glazed panels. The cables are routed though the mullion. (From Roberts, S., and Guariento, N., *Building Integrated Photovoltaics: A Handbook*, Walter de Gruyter GmbH, 2009.)

(a) (b)

FIGURE 5.10 (a) Curtain wall equipped with relatively small crystalline silicon cells with large spacing between the cells (Balenciaga store, Miami, FL, USA) (From Onyx Solar, https://www.onyxsolar.com/balenciaga-store-miami, 2019a.); and (b) photograph of the 1 m^2 Electric Mondrian™ with the Utrecht Dom tower in the back. (From van Sark, W. et al., *Solar RRL*, 1, 1600015, 2017.)

thin-film technologies, such as amorphous silicon or organic PV, are used, transparency is defined by the density of the PV material. Figure 5.12 shows a skylight system with different transparencies, with a peak power of 28, 34, and 40 Wp/m^2 (Onyx Solar 2019b). In both crystalline and thin-film technologies, the energy yield is inversely proportional to the transparency.

An upcoming technology in the field of solar glazing is luminescent solar concentrators (LSC's) as illustrated in Figure 5.10b. In this case, windows are colored with a dye that guides a specific part of the incoming photons to the sides of the glass. On these sides, solar cells are attached that are capable of generating electricity out of the diverted photons with a high efficiency. Even though the total efficiency of such a system is still rather low (1%), the ability to vary with different aspects such as color, concentration (transparency), and size offers design features that are desirable in the field of BIPV (van Sark et al. 2017).

5.3.2 Skylight Systems

A skylight system or canopy generally has a semi-transparent character that allows natural light to enter the space underneath. The system must also be waterproof and be able to deal with the external climatic elements. Construction of glazed skylights can

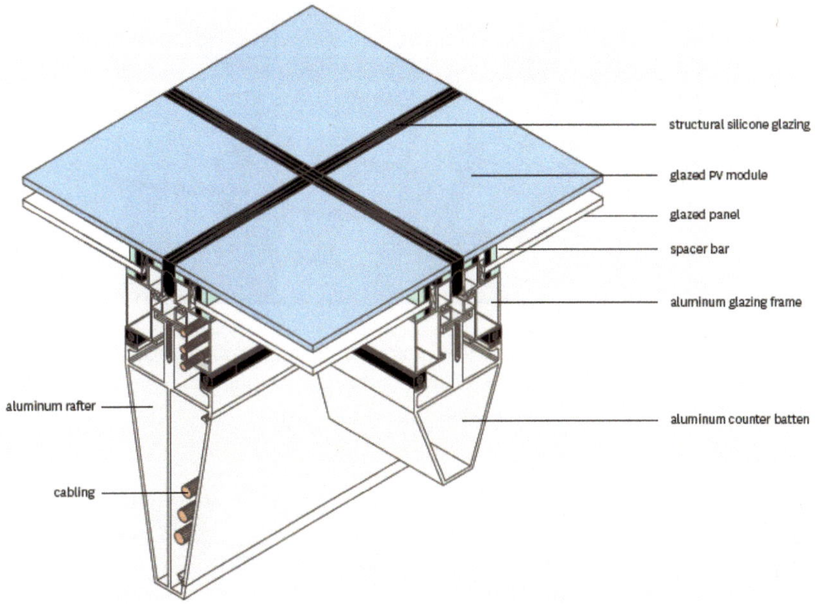

FIGURE 5.11 Detail of a horizontal overhead glazing system with PV modules. (From Roberts, S., and Guariento, N., *Building Integrated Photovoltaics: A Handbook*, Walter de Gruyter GmbH, 2009.)

be based on stick curtain wall systems and unitized curtain wall systems. The double-glazed units can be replaced by clear or opaque PV modules, preferably double glazed. An additional outer layer can be applied for solar control, high-performance coatings, and low emissivity. Figure 5.11 shows a detail of a horizontal overhead glazing system. The vertical beams between glass units (mullions) act as rafters (sloping beam supporting a roof) in the structure, and the horizontal beams between glass units (transoms) act as counterbalance to transfer vertical loads (dead load, wind, snow, maintenance) to the main structure. Cables are hidden in the cavities of the rafters.

Figure 5.12 shows a part of the skylight system of the refurbished historic food market of Bejar in, Salamanca (Spain). Amorphous silicon modules with varying degrees of transparency (and peak power) are combined with colored glass. The installed capacity of the whole system is 6.7 kWp (Onyx Solar 2019b).

5.3.3 Roof Systems

A **full roof solution** is shown in Figure 5.13, where the key points clearly are a homogeneous and integrated aesthetically pleasant appearance combined with electricity production and protection against weather conditions. The system in the figure is a renovation project in Maastricht (The Netherlands), which consists of 33 monocrystalline silicon solar panels and 9 dummy panels around the skylight and the chimney. Costs are around 300 €/m^2 with a power production of 150 up to 190 Wp/m^2 (Beausolar 2019).

FIGURE 5.12 Skylight on the historic food market of Bejar, Salamanca (Spain). Amorphous silicon modules with varying degrees of transparency are combined. (From Onyx https://www.onyxsolar.com/bejar-market, 2019b.)

FIGURE 5.13 House renovated with a full roof BIPV solution of Beausolar. Around the chimney and skylight dummy pieces are installed to retain a homogeneous appearance. (From Beausolar, Beausolar Project Webpage. https://beausolar.eu/projecten/, 2019.)

For buildings where no full roof system is required, different combinations with conventional roofing systems are available. An example of an **in-roof system with standard-sized solar panels** is the Clearline Fusion of Viridian Solar. Figure 5.14 shows how these modules are combined with standard curved roof tiles. The solar modules measure 992 × 1640 mm and provide a peak power of 260 up to 300 W. The adjusted frame allows it to be fixed on the existing tile battens with dedicated brackets. Around the panels flashings are placed to guide the water (Viridian Solar 2017). A schematic drawing of the attachment system is shown in Figure 5.15.

FIGURE 5.14 A house in the north of the Netherlands, equipped with Viridian Solar's Clearline Fusion. An in-roof PV system with standard-sized solar modules.

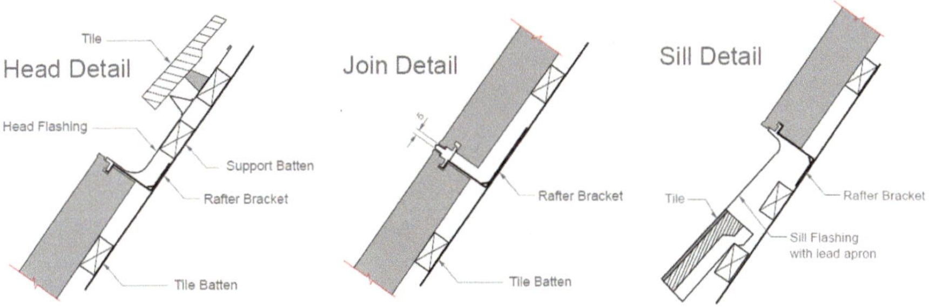

FIGURE 5.15 Schematic drawing of how the Clearline Fusion is built into the conventional roof system. (From Viridian Solar, Clearline Fusion PV16 datasheet, http://www.viridiansolar.co.uk/assets/files/Clearline-fusion-PV16-Data-Sheet.pdf, 2017.)

Another example of a BIPV roof system that combines with conventional tiles is the Solinso Mystiek **solar roof tile** (Figures 5.1a and 5.16). These relatively large tiles have a visible length of 1500 mm and a width of 420 mm, generating a peak power of 85 W (Solinso 2018a). The advantage of this system is that the panels blend in with the rest of the roof, while minimizing the amount of connections to be made during installation. The yield per square meter of this system is similar to standard PV modules. Figure 5.16 shows how these panels are placed on a roof in combination with traditional flat roof tiles.

Curved roof tiles are especially developed for cases where flat roof elements, such as described above, still do not meet the aesthetical requirements for a specific site, for example, on monumental buildings where the roof should keep the appearance of the

FIGURE 5.16 Solinso Mystiek solar roof tile combined with skylights and standard flat rooftiles. (From Solinso, Solinso brochure 2018: Mystiek Zonnedakpannen, https://www.solinso.nl/images/downloads/Brochure-Solinso-2018-.pdf., 2018b.)

classic curved rooftiles it used to have. Figure 5.17a illustrates how two solar cells are integrated into a ceramic tile that also maintains its traditional red color. This product provides a peak power of almost 80 W/m^2 at a price of 256 €/m^2. The all-black counterpart of this tile has a peak power of 90 W/m^2 and costs 205 €/m^2, showing that the loss in efficiency for being red instead of black is only 11% (Zep 2019b).

5.3.4 Shading Systems

The comfort of building occupants is improved with respect to daylight and views to the outside by introducing large vision areas in façades. A passive way to limit the effect of solar gains due to these large vision areas is the application of passive shading devices. Since these shading devices naturally receive large amounts of sunlight, the integration of solar cells into such systems is a rational measure.

An example where BIPV is utilized in a shading system is the ECN office building 31 in Petten, the Netherlands (Figure 5.18). Overheating was one of the main issues of this building, and either an improved air-conditioning system or a shading device was

(a) (b)

FIGURE 5.17 (a) Application of ceramic solar roof tiles on the roof of a house in the Netherlands; 1 m^2 of this roof holds 9.8 tiles, providing a peak power of almost 80 Wp/m^2 (From Zep, Zep Website, Reference Residence Tiel, https://www.zep.solar/nl/referenties/woning-tiel, 2019c.); and (b) ceramic red roof tile with two red-colored solar cells integrated on its surface, providing a peak power of 8 Wp per tile. (Zep, Datasheet redline, https://www.zep.solar/images/downloads/datasheet-Redline_pdf_1.pdf, 2019a.)

(a) (b) (c)

FIGURE 5.18 (a) BIPV shading system installed on ECN office building 31, (b) front side sketch, (c) schematic drawing of the mounting system of the shading lamellas. (BEAR architects, http://bear-id.com/projects/ecn-building-42/, 2019.)

required. The latter option has been selected, combined with BIPV installed on the roof and integrated into the shading system. The BIPV system on the south façade consists of a mounting system managed by software. This software takes into account both the optimum tilt of the system and the internal lighting, finding the perfect relationship between both phenomena. However, when preferred, the tilt of the shading system can be manually managed. Each lamella is 840 mm wide and has a length of 3 m. On top of the aluminum lamella three PV modules (Shell Solar type RSM 50) are installed with a peak production of 48 Wp. In total, the south façade has a PV-system power of 26.21 kWp. The roof-system is installed as a parasol, which covers an active cooling system by ventilators and air ducts. Underneath this cooling layer a rock–wool layer is installed to ensure proper insulation (Prasad and Snow 2005).

Another, more extraordinary, shading system is the Lumiduct™ of Wellsun. It blocks direct sunlight and converts it into electricity, while maintaining the majority of incoming diffuse light to enter the building. The system consists of an array of panels that are located between a glass double-skin façade with a width of 1 m. Each panel is equipped with PV solar concentrators, which track the movement of the sun and have a peak power of 300 W/m^2 (Figure 5.19). The company claims that the electricity generation of the system is similar to that of a conventional PV system mounted on a façade (with 70% of the surface covered with PV). An extra feature is the possibility to add LEDs to the panels, which allow the system to work as a media wall at night (Figure 5.20) (Wellsun 2019).

(a) (b)

FIGURE 5.19 a) First full scale project of Wellson Lumiduct™ installed in a façade, and (b) closeup of the panels from inside the building (Mondial Movers, Alblasserdam, NL).

(a) (b)

FIGURE 5.20 (a) Artist impression of the Lumiduct™ system by day, and (b) by night. (Designed for Havenbedrijf Rotterdam, Merwe-Vierhaven area).

5.4 BALANCE OF SYSTEM AND MOUNTING SYSTEMS

5.4.1 Inverters or Converters

Just as with any other PV application, in a BIPV system the solar cells produce electricity in a form of direct current (DC) with a varying current and voltage. To be able to make use of this power with electric appliances, the electricity has to be shaped in the right form.

Inverters invert the DC current into an alternating current (AC) with a frequency and a voltage that matches the local grid (e.g., 230 V and 50 Hz). Many appliances that are connected to this grid, subsequently, convert AC back to DC before making use of the electricity. To omit these steps and the losses that it encompasses, there are examples where the DC current is kept. In these cases, a **DC-DC converter** converts the varying DC voltage of the PV system into a voltage that matches the system it powers. Examples are cases where the BIPV system directly feeds a battery or an electric vehicle. A more comprehensive application is a BIPV system that feeds a local (building) DC grid, where appliances such as lighting, computers, televisions, or even the HVAC systems are connected to.

5.4.2 Optimizing Energy Yield

To optimize the yield of the BIPV system, maximum power point trackers (MPPT) are utilized. MPPTs are generally part of the inverter; they adapt the load (as seen from the solar panels) such that an electrical operating condition is obtained, which maximizes the energy yield of the PV system. A string of several solar panels is generally connected in series to such an MPPT. To optimize the operation of the MPPT and the total yield of the string, the physical operating conditions (irradiation, shading) of each PV module inside the string should be similar. Especially in the case of BIPV, where shading of solar panels is common, but also relatively predictable, it is very important to make well-thought decisions on which solar panels are combined in each string. Instead of connecting strings of multiple modules to one central inverter, **microinverters** are also available. In this case, each solar panel has its own inverter, which allows it to operate at its own maximum power point, each panel is directly connected to the AC grid and not influenced by other panels in the system. Another way to circumvent the influence of adjacent modules in each string is the use of **power optimizers**. These devices are placed behind each solar panel and maintain operation at the optimum power point of the panel. Strings of panels are still connected in series with a central inverter, which inverts the DC current into AC. Especially in BIPV façades, where shading is more of an issue than on roofs, the use of power optimizers has benefits. Not only because it enhances the ability to cope with shading, but also because it is a way to monitor each BIPV module individually. Since PV modules on façades are generally harder to reach physically, this facilitates inspection, maintenance, or replacement of modules.

5.4.3 Mounting Systems

A common mounting procedure for BAPV is placing an array of extruded aluminum profiles on which the solar panels are being attached with bolts or clamps. Similar methods are applicable for BIPV solutions where the main function of the solar panels (besides energy generation) is aesthetic appearance, for example, on refurbished buildings where insulation, weathertightness, and structural strength are provided by the existing wall behind the solar panels, and where a new outer skin of (colored) solar panels gives the building a new appearance.

However, as earlier described in this chapter and as the overview of the different types of BIPV in Figure 5.3 suggests, BIPV generally fulfils more functions and is more deeply integrated into the building skin. In these cases, other mounting systems are required, which have more similarities with the systems that are used for building elements, which are replaced by the BIPV system. Figures 5.6 and 5.9 show examples of these systems. An important adaptation to the original non-BIPV system to meet BIPV needs is accommodating the electric connection between the solar panels. The dedicated spaces for cable guidance in Figures 5.6 and 5.9 are examples of such an adaptation.

This electric connection is in fact a feature that made the development of a European standard for BIPV elements a challenging exercise. By its nature, a BIPV element is both a construction element and an electrical device. The formation of the EN 50583 "Photovoltaics in Buildings" standard (published in 2016) required a linkage between two separated areas in the standardization world, connecting the field of building construction (CEN) with the field electrical devices and systems (Erban 2016; CENELEC 2016a, 2016b)

5.5 ENERGY PERFORMANCE OF BIPV

As described in the preceding paragraph, shading is an important factor for the performance of BIPV systems. Beside these, shading effects, location, orientation, and tilt (for façades generally 90 degrees) of the system must be taken into account. In the subsequent sections, the general features that are typical for the yield of BIPV systems are illustrated with two case studies.

5.5.1 BIPV Façade ZSW Laboratory and Office Building

The Zentrum für Sonnenenergie- und Wasserstoff-Forschung Baden-Württemberg (ZSW) laboratory and office building at the Stuttgart Engineering Park (STEP) is realized with a ventilated rainscreen façade, where 256 m^2 of the opaque façade area of the highest cuboid of the building is equipped with thin-film (CIGS) BIPV façade elements. These elements are placed at the southeast (SE), southwest (SW), and northwest (NW) side of the building. Further on, an open-rack PV system is placed on the roof, tilted 40°, and oriented south (S40). A picture of the building is shown in Figure 5.21 (Geyer et al. 2018).

Table 5.2 shows the specific yield of individual PV modules on different façades and on the roof. These yields are based upon a dataset from July 2017 to June 2018. The unshaded module on the southwest façade (SW1) has a yield of 75% compared to the open-rack module. This illustrates that, even though the tilt of a façade is not optimal, yields are still significant. The yield on the northwest façade is less than 40% of the optimum, showing how orientation plays a very important role in the

FIGURE 5.21 PV Façade at the ZSW laboratory and office building in Stuttgart (Germany). The green line indicates the building block on which BIPV elements are installed. (From Geyer, D et al., Research project CIGS-Façade, PV façades—chances and limits, *Proceedings of the 35th European Photovoltaic Solar Energy Conference*, pp. 1805–1808, 2018.)

TABLE 5.2

Comparison of Specific Yield (July 2017–June 2018) of Individual MPP-Tracked PV Modules within the SE-, SW-, NW-Façade and a Rack-Mounted Module Facing South at 40 Tilt

	SE3 Façade Moderate Shading	SE4 Façade Heavy Shading	SW1 Façade Shade-free	NW6 Façade	S40 Open Rack
Specific yield (kWh/kWp)	659	497	749	380	999
Relative to S40° (%)	66	50	75	38	100

Source: Geyer, D., et al., Research project CIGS-Façade, PV façades—chances and limits, *Proceedings of the 35th European Photovoltaic Solar Energy Conference*, pp. 1805–1808, 2018.

energy yield of a BIPV installation. Further on, the effect of shading is very clear, SE3 experiences shading from a higher building on the opposite side of the street in the east. SE4 is shaded by an additional building edge of its own façade. The heavy shading on SE4 reduces the yield to 50% of the open-rack system, where a similar yield as SW1 could be expected in a shade-free situation.

Figure 5.22 shows how the specific power for each façade and for the south-oriented open rack module varies during June 20, 2018. In the morning (when the sun rises in the east), the modules on the southeast side of the building show their main production. The open-rack system peaks around noon, and the southwest façade has its highest production in the afternoon. The northwest façade peaks at the start of the evening.

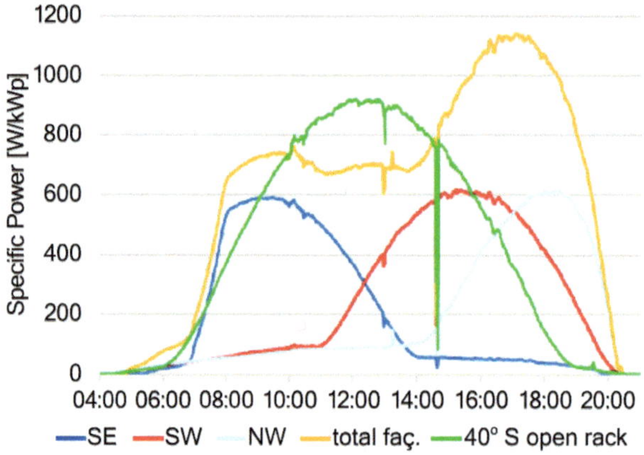

FIGURE 5.22 Course of specific power output on June 20, 2018 for the façade modules and for the open-rack module. (From Geyer, D et al., Research project CIGS-Façade, PV façades—chances and limits, *Proceedings of the 35th European Photovoltaic Solar Energy Conference*, pp. 1805–1808, 2018.)

Compared to a system that is only oriented south, a BIPV system installed on façades with different orientations causes a more evenly distributed energy yield during the day. This generally corresponds more to the energy demand of the building, which enhances its degree of self-consumption, and as a consequence this could ultimately lead to significant savings in electricity storage or fossil fuel-based backup power (Hummon et al. 2013).

5.5.2 BIPV Potential on Larger Scale in Lisbon

The example of the ZSW building in Germany showed an annual specific yield of nearly 1000 kWh/kWp for the PV system on the roof, where the unshaded BIPV façade on the southeast side produces about 750 kWh/kWp each year. It also illustrates how shading from other buildings influences the energy production. A simulation for two residential areas of Lisbon (Portugal), carried out by Brito et al. (2017), results in the specific yield for each façade orientation, and up to which level this yield accommodates the electricity demand. Figure 5.23 presents the annual irradiation on both areas. The irradiation on the roofs is significantly higher than on the façades, and the effect of shading on façades from adjacent buildings is clearly visible. Figure 5.24 displays a histogram of the solar potential of the areas. It confirms how the roofs have an irradiation from 1000 to 1800 kWh/m^2, where façades reach levels between 100 and 1000 kWh/m^2. The south-facing façades are clearly at the higher end of the spectrum, with east and west in the middle and north in the lower regions. Important to note is the amount of available façade area compared to the amount of roof area.

Figure 5.25 illustrates how the combination of irradiation and façade area leads to a monthly potential PV yield. The roofs still take up the largest part of the energy

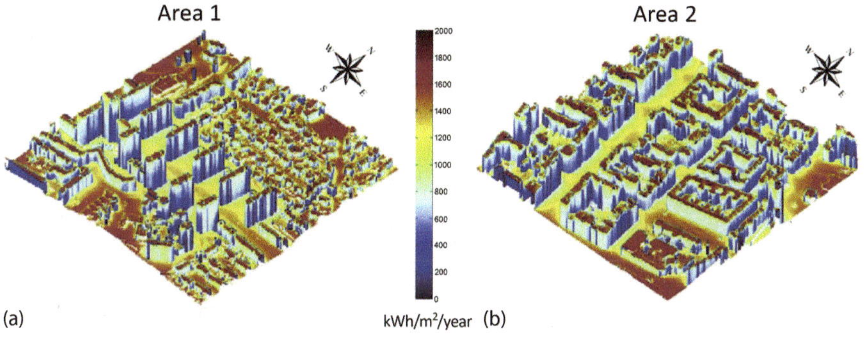

FIGURE 5.23 Annual solar irradiation for (a) Area 1, and (b) Area 2. (From Brito, M.C. et al., *Renewable Energy*, 111, 85–94, 2017.)

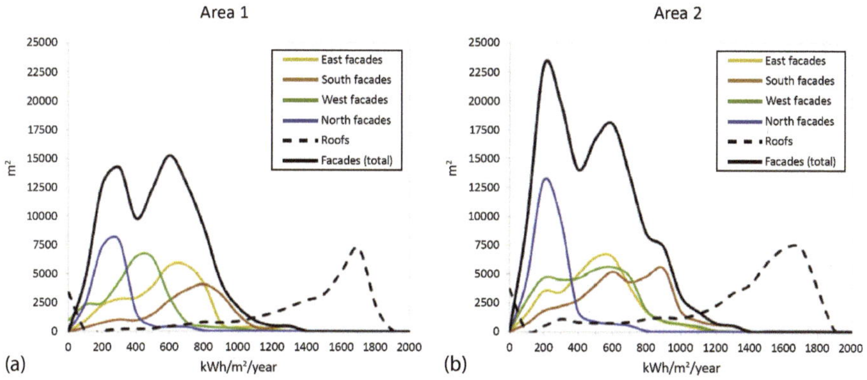

FIGURE 5.24 Annual solar potential histogram for (a) Area 1, and (b) Area 2. (From Brito, M.C. et al., *Renewable Energy*, 111, 85–94, 2017.)

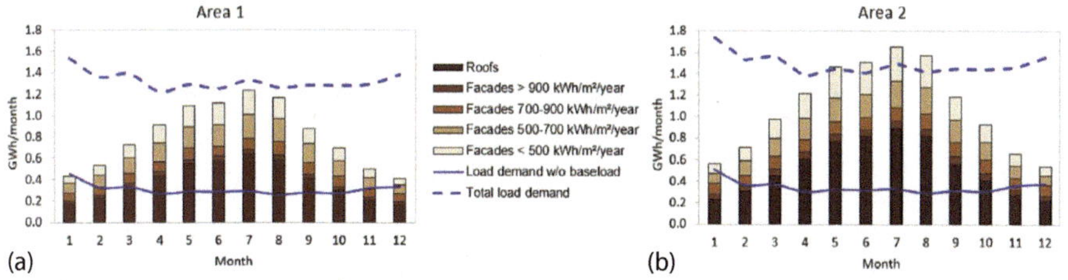

FIGURE 5.25 Monthly PV potential for (a) Area 1, and (b) Area 2. (From Brito, M.C. et al., *Renewable Energy*, 111, 85–94, 2017.)

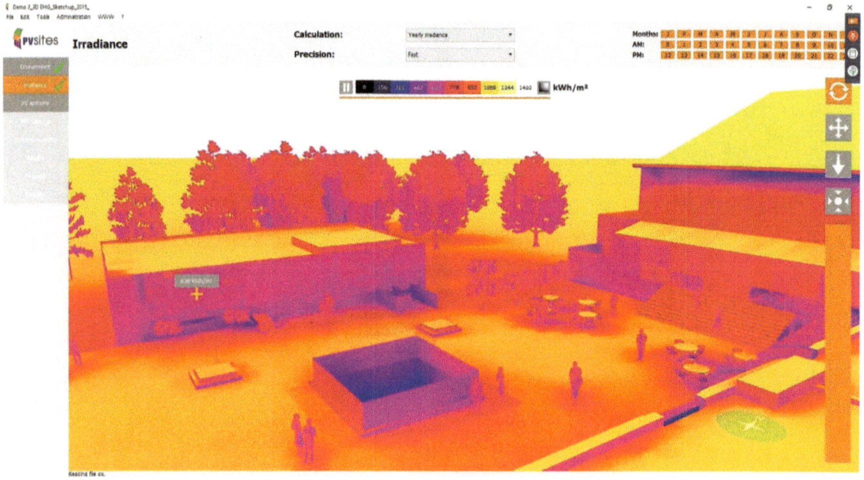

FIGURE 5.26 Screenshot of PVSites simulation software. The different colors visualize the yearly irradiance on each surface of the modeled scenery. (From PVSites, Homepage of PVSites Software Package, https://www.pvsites.eu., 2019.)

yield, but the façades all together are able to generate almost the same amount of energy. In the summer months, Area 2 is able to satisfy all of its energy demand. Both areas are able to satisfy their load demand without baseload all throughout the year when full capacity is utilized.

5.5.3 Simulation Software

To support designers in optimizing their choice of PV modules, inverter types, and string configuration, different software packages are available. With these programs, BIPV systems can be designed and simulated such that the irradiation and the occurrence of shading by the building itself or its surroundings is predicted. Figure 5.26 gives a screenshot of the PVSites simulation software. A 3D model of a building, its surroundings, and its orientation can be read by this software package. The combination of the model and local irradiation data provides the designer with a detailed visualization of the irradiance on site. Besides these visualizations, a (BI)PV system can be designed with the software, and revenues are predicted. This includes designing strings and inverter sizing to optimally manage shading effects on performance (PVSites 2019).

5.6 COST ASPECTS

A price survey was performed in 2017 by SUPSI and SEAC, amongst a total of 35 participants mainly in the Netherlands and Switzerland. It provides an insight in the price for BIPV roof and façade elements in €/m² (Figures 5.27 and 5.28). The costs are compared with the costs for conventional applications, from which the data is obtained from Swiss building databases. Even though the number of participants on the façade side (seven) is too small to generalize results, the figures provide an interesting overview of

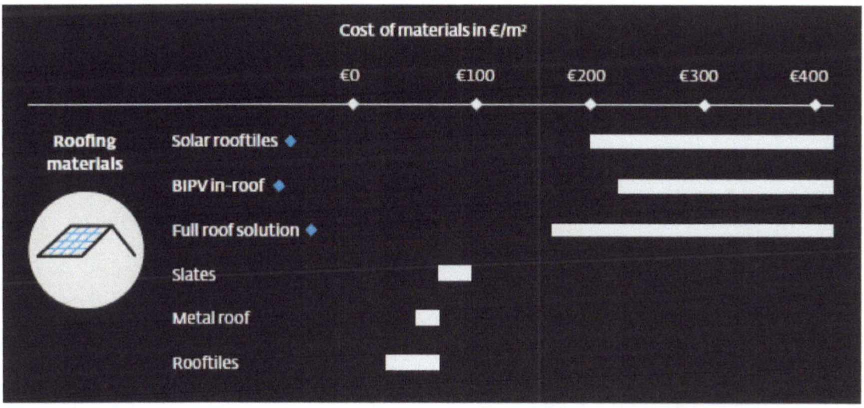

FIGURE 5.27 Cost of BIPV roof tiling compared to other roofing materials used in conventional pitched roof. (From Zanetti, I. et al., Building integrated photovoltaics: Product overview for solar building skins, Eindhoven (NL) & Lugano (CH), www.seac.cc, 2017.)

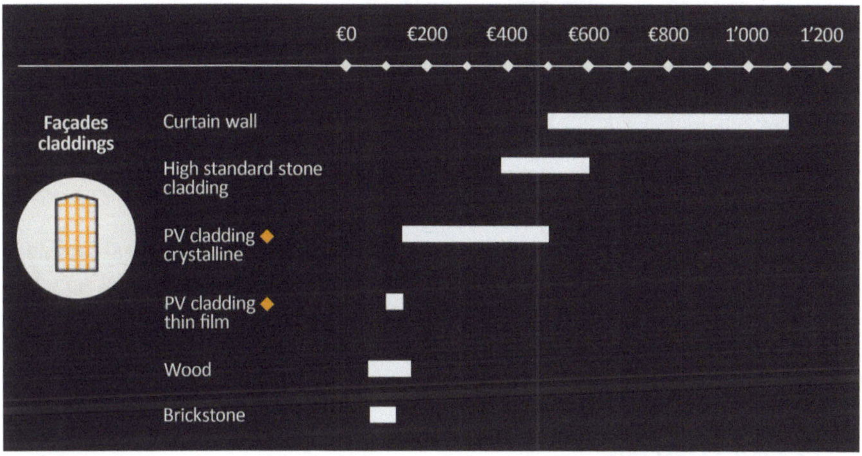

FIGURE 5.28 PV costs compared to other cladding materials used in the built environment as façade cladding materials. (From Zanetti, I. et al., Building integrated photovoltaics: Product overview for solar building skins, Eindhoven (NL) & Lugano (CH), www.seac.cc, 2017.)

the costs (Zanetti et al. 2017). Figure 5.28 shows, for example, that PV cladding has lower initial costs than high-standard stone cladding. These figures only represent the costs for purchasing and installing the materials. When considering BIPV, the costs of the building materials that are replaced can be subtracted from the cost of the BIPV materials. Contrary to conventional building materials, PV systems also generate electricity, thus providing a revenue from the investments during their lifetime. The figures show clearly that the choice for BIPV is not only based on economic motives.

The costs of the BIPV system on the ZSW building from Section 5.5.1 are illustrated in Figure 5.29. It shows that the PV façade system cost amounts to 341 €/m^2 on

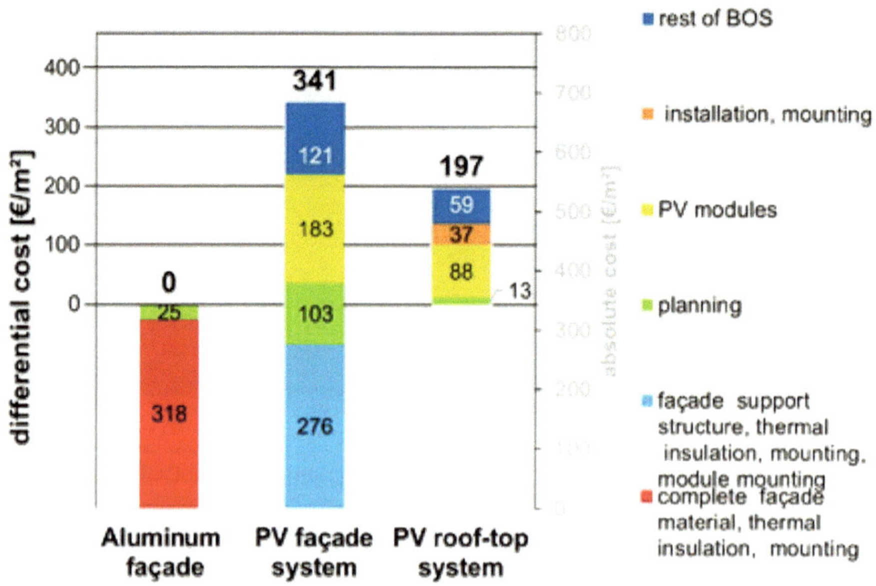

FIGURE 5.29 Comparison of area-related costs of the aluminum and PV-façade system (as built). Costs for a standard rooftop PV system are shown for comparison. (From Geyer, D et al., Research project CIGS-Façade, PV façades—chances and limits, *Proceedings of the 35th European Photovoltaic Solar Energy Conference*, pp. 1805–1808, 2018.)

top of the costs of the aluminum façade without PV. Due to architectural demands, six different module formats where used in this project, resulting in a situation where only 27% of all modules could be realized as standard CIGS modules. A further analysis of the module costs revealed that any deviation from standard module sizes increased the price 2.5 up to 3 times per unit area.

5.7 CONCLUSIONS

This chapter has illustrated that BIPV is a technology with a great future. Not only is the expected market size to increase about fivefold in the next 10 years, also a myriad of technological solutions have been developed, and there is more to come. An important driver is the need to develop NZEBs for which on-site generation of renewable energy is highly desirable. This requires improved buildings designs foremost, while BIPV can play a crucial role in generating energy from the roof and façades, either opaque or transparent. This is particularly of importance for high-rise buildings and environments. As the chapter's examples clearly show, project developers and architects have a lot of freedom to design aesthetically pleasing renewable energy solutions, thus helping to design sustainable cities of the future.

REFERENCES

Ballif, C., L.-E. Perret-Aebi, S. Lufkin, and E. Rey. (2018). Integrated thinking for photovoltaics in buildings. *Nature Energy* 3: 438–342.

Beausolar. (2019). Beausolar Project Webpage. https://beausolar.eu/projecten/.

BEAR architects. (2019). R&D building ECN Petten. http://bear-id.com/projects/ecn-building-42/.

Biyik, E., M. Araz, A. Hepbasli, M. Shahrestani, R. Yao, L. Shao, E. Essah et al. (2017). A key review of building integrated photovoltaic (BIPV) systems. *Engineering Science and Technology, an International Journal* 20: 833–858.

Breyer, C., D. Bogdanov, A. Gulagi, A. Aghahosseini, L.S.N.S. Barbosa, O. Koskinen, M. Barasa et al. (2017). On the role of solar photovoltaics in global energy transition scenarios. *Progress in Photovoltaics: Research and Applications* 25: 725–745. doi:10.1002/pip.2885.

Brito, M.C., S. Freitas, S. Guimaraes, C. Catita, and P. Redweik. (2017). The importance of facades for the solar PV potential of a mediterranean city using LiDAR data. *Renewable Energy* 111: 85–94. doi:10.1016/j.renene.2017.03.085.

CENELEC. (2016a). *EN 50583-1.*

CENELEC. (2016b). *En 50583-2.*

Clua Longas, A., S. Lufkin, and E. Rey. (2017). Towards advanced active façades: Analysis of façade requirements and development of an innovative construction system. In *PLEA 2017, Edinburgh, Design to Thrive.* Hongkong: PLEA.

Defaix, P.R., W.G.J.H.M. van Sark, E. Worrell, and E. de Visser. (2012). Technical potential for photovoltaics on buildings in the EU-27. *Solar Energy* 86(9). doi:10.1016/j.solener.2012.06.007.

Delponte, E., F. Marchi, F. Frontini, C. Polo, K. Fath, and M. Batey. (2015). BIPV in EU28, from niche to mass market: An assessment of current projects and the potential for growth through product innovation. In *31st European Photovoltaic Solar Energy Conference and Exhibition*, pp. 3046–3050.

Efurosibina Attoye, D., K.A. Tabet Aoul, and A. Hassan. (2017). A review on building integrated photovoltaic façade customization potentials. *Sustainability* 9: 2287. doi:10.3390/su9122287.

Erban, C. (2016). *BIPV Standard EN 50583 »Photovoltaics In Buildings« Released.*

European Commission. (2010). Directive 2010/31/EU of the European Parliament and of the Council of 19 May 2010 on the energy performance of buildings. *Official Journal of the European Union* 18(6): 13–35.

European Commission. (2016). Clean Energy for All Europeans, Communication from the Commission to the European Parliament, the Council, the European Economic and Social Committee, the Committee of the Regions and the European Investment Bank. COM (2016) 860 Final.

Farkas, K., L. Maturi, A. Scognamiglio, F. Frontini, M. Cristina, M. Probst, C. Roecker, and M. Lundgren. (2015). Designing Photovoltaic Systems for Architectural Integration, Criteria and Guidelines for Product and System Developers, Report T.41.A.3/2: IEA SHC Task 41 Solar Energy and Architecture.

Ferrara, C., T. Kuhn, H. Wilson, and R. Schindler. (2012). BIPV–From Add-On PV to Real Integration Challenges and Options. Keynote presentation at 2012 International Conference on Building Envelope Systems and Technologies. https://cdn2.hubspot.net/hub/55819/file-14734737- pdf/docs/bipv-keynote-icbest.pdf.

Ferrara, C., H.R. Wilson, and W. Sprenger. (2017). Building-Integrated Photovoltaics (BIPV). In *The Performance of Photovoltaic Systems*, edited by N. Pearson, pp. 235–250. Amsterdam, the Netherlands: Elsevier.

Flexipol. (2019). Workshop voor architecten: BIPV een spectrum aan mogelijkheden. https://www.bipvnederland.nl/workshop-voor-architecten-bipv-een-spectrum-aan-mogelijkheden/.

Frontini, F., P. Bonomo, A. Chatzipanagi, G. Verberne, M. Van den Donker, K. Sinapis, and W. Folkerts. (2015). BIPV product overview for solar façades and roofs—BIPV status report 2015. Eindhoven (NL) & Lugano (CH).

Geyer, D., P. Lechner, S. Hummel, D. Stellbogen, and C. Conejo Gangkofner. (2018). Research project CIGS-Façade, PV façades—chances and limits. *Proceedings of the 35th European Photovoltaic Solar Energy Conference*, pp. 1805–1808.

Global Industry Analysts. (2015). The global building integrated photovoltaics (BIPV) market.

Heinstein, P., C. Ballif, and L.-E. Perret-Aebi. (2013). Building integrated photovoltaics (BIPV): Review, potentials, barriers and myths. *Green* 3(2): 125–156. doi:10.1515/green-2013-0020.

Hummon, M., P. Denholm, and R. Margolis. (2013). Impact of photovoltaic orientation on its relative economic value in wholesale energy markets. *Progress in Photovoltaics: Research and Applications* 21(7): 1531–1540. doi:10.1002/pip.2198.

IEA PVPS. (2019). 2019 Snapshot of global photovoltaic markets. *IEA PVPS T1-35:2019.*

Jelle, B.P., C. Breivik, and H.D. Røkenes. (2012). Building integrated photovoltaic products: A state-of-the-art review and future research opportunities. *Solar Energy Materials and Solar Cells* 100: 69–96. doi:10.1016/j.solmat.2011.12.016.

Kameleonsolar. (2019). Kameleonsolar Company Web Site. https://www.kameleonsolar.com.

Krawietz, S., J. Poortmans, G. Masson, F. Pause, J. Palm, and R. Schlatmann. (2016). Building integrated photovoltaics (BIPV) as a core element for smart cities–integrated research, innovation and competitiveness strategies.

Kuijpers. (2019). Kuijpers Company Web Site. https://www.kuijpers.nl.

Lee, J B, J.W. Park, J.H. Yoon, N.C. Baek, D.K. Kim, and U.C. Shin. (2014). An empirical study of performance characteristics of BIPV (Building Integrated Photovoltaic) system for the realization of zero energy building. *Energy* 66: 25–34. doi:10.1016/j.energy.2013.08.012.

n-tech Research. (2018). BIPV market forecast and analysis 2018–2027. https://www.ntechresearch.com/market-reports/bipv-market-forecast/.

Onyx Solar. (2019a). Balenciaga Miami, curtain wall. https://www.onyxsolar.com/balenciaga-store-miami.

Onyx Solar. (2019b). https://www.onyxsolar.com/bejar-market.

Osseweijer, F.J.W., L.B.P. Van den Hurk, E.J.H.M. Teunissen, and W.G.J.H.M. van Sark. (2018). A comparative review of building integrated photovoltaics ecosystems in selected European countries. *Renewable and Sustainable Energy Reviews* 90: 1027–1040.

Prasad, D.K., and M.D. Snow. (2005). *Designing with Solar Power: A Source Book for Building Integrated Photovoltaics.* London, The Images Pub. Victoria, Australia: Images Publishing Group.

PVSites. (2019). Homepage of PVSites Software Package. https://www.pvsites.eu.

Ritzen, M., Z. Vroon, C. Geurts. (2017). Building integrated photovoltaics. In *Photovoltaic Solar Energy: From Fundamentals to Applications*, edited by A. Reinders, P. Verlinden, W. Van Sark, and A. Freundlich, pp. 579–589. Chichester, UK: Wiley.

Roberts, S., and N. Guariento. (2009). *Building Integrated Photovoltaics: A Handbook.* Basel: Walter de Gruyter GmbH.

Scognamiglio, A. (2017). Building-integrated photovoltaics (BIPV) for cost-effective energy-efficient retrofitting. *Cost-Effective Energy-Efficient Building Retrofitting* 169–197. doi:10.1016/B978-0-08-101128-7.00006-X.

Shukla, A.K., K. Sudhakar, P. Baredar, and R. Mamat. (2017). BIPV in Southeast Asian countries—Opportunities and challenges. *Renewable Energy Focus* 21: 25–32. doi:10.1016/j.ref.2017.07.001.

Solinso. (2018a). Datasheet Mystiek. https://www.solinso.nl/images/downloads/20180513_Datasheet-Mystiek-85Wp.pdf.

Solinso. (2018b). Solinso Brochure (2018): Mystiek Zonnedakpannen. https://www.solinso.nl/images/downloads/Brochure-Solinso-2018-pdf.

Sorba. (2019). Sorba Company Web Site. https://www.sorba.com.

Studio Solarix. (2019). Studio Solarix Company Web Site. https://www.studio-solarix.nl.

Tripathy, M., P.K. Sadhu, and S.K. Panda. (2016). A critical review on building integrated photovoltaic products and their applications. *Renewable and Sustainable Energy Reviews* 61: 451–465. doi:10.1016/j.rser.2016.04.008.

van Sark, W., P. Moraitis, C. Aalberts, M. Drent, T. Grasso, Y. L'Ortije, M. Visschers, M. Westra, R. Plas, and W. Planje. (2017). The "Electric Mondrian" as a luminescent solar concentrator demonstrator case study. *Solar RRL* 1(3–4): 1600015. doi:10.1002/solr.201600015.

Viridian Solar. (2017). Clearline fusion PV16 datasheet. http://www.viridiansolar.co.uk/assets/files/Clearline-fusion-PV16-Data-Sheet.pdf.

Wellsun. (2019). Lumiduct Brochure. https://docs.wixstatic.com/ugd/c50ab0_896434ed99ec4cc2add73bfff3861ee5.pdf.

Zanetti, I., P. Bonomo, F. Frontini, E. Saretta, M. van den Donker, F. Vossen, and W. Folkerts. (2017). Building integrated photovoltaics: Product overview for solar building skins. Eindhoven (NL) & Lugano (CH). www.seac.cc.

Zep. (2019a). Datasheet Redline. https://www.zep.solar/images/downloads/datasheet-Redline_pdf_1.pdf.

Zep. (2019b). Zep Product Website. https://www.zep.solar/en/products.

Zep. (2019c). Zep Website, Reference Residence Tiel. https://www.zep.solar/nl/referenties/woning-tiel.

Design Case 5

The Solaris Building

Alessandra Scognamiglio[1]

5.1 INTRODUCTION

For a long time, the use of solar technologies has been seen with suspicion by architects because of their *visibility*.

Any building technology, or material choice, has an influence on the aesthetics of the architectural system in which it is used, and some technologies, that is, photovoltaics (PVs), due to their high recognizability, are even more visible than others.

The appearance of PVs, for many years has been limited to a small variety of components, mainly designed for getting the maximum efficiency, given the high cost of the cells. This translated into square, blue, opaque modules, in a small variety of sizes, which would significantly influence the visual aspects of a building in which they would be integrated.

This "visibility" could be an intentional choice in terms of "style," when the main message of a building is something like, "I generate my own energy"(Scognamiglio and Privato 2008; Scognamiglio et al. 2014). Nevertheless, in many other cases, where the intention of certain architecture is not directly related to such kind of message, the presence of PVs can be *intrusive*.

For these reasons, very often architects refused—and still refuse—the use of PVs. A building with PVs is in a way "featured" as a solar building, with a probable damage for the other meanings that a building can entail.

With the time passing, the fast decrease in the cost of PV cells allowed an increased freedom in designing PV modules suitable for architecture. The main point for the modules' design was not their mere efficiency anymore, but rather the match with some visual requirements given by the architects. A typical one is that "Photovoltaics should be invisible; nobody should know that this is a solar building."

Actually, this was made possible by the introduction on the market of some technologies that advantage the choice of color and texture of the surface of the module, while keeping the efficiency under control. Efficiency losses, which were not justifiable some years ago, are nowadays acceptable thanks to the low price of solar cells.

Therefore, among other new products, glass panels of many sizes, colors, and textures popped up on the market, creating an opportunity window for the architects' wishes.

[1] This chapter is the author's elaboration based on: Caspar Schärer, Un pezzo forte, in Solaris #1, 2018, pp. 3–17.

In this specific approach, the PV part of the module is quite traditional; it is a glass–glass module (made of three layers: glass–PV cells–glass). But the front glass side is the space for freedom. Here it is possible to experiment with any possibility offered by glass printing techniques, and, also with the introduction of additional layers that can change the texture and the pattern of the component. With the use of glass–glass PV modules, where the architects can "play around" the external layer, so as to get the image they like, a new domain for architectural experimentation is opened.

This change in perspective, opens a new scenario, where the architects can finally give their contribution to an epochal demand: How does architecture include net-zero energy features in its program of demands and wishes, without sacrificing any of its traditional meanings and disciplinary tools?

5.2 THE SOLARIS BUILDING IN ZÜRICH, A TWIST IN THE STORY

The building presented in this design case is an emblematic example for the change in perspective described above: PVs, from being an *intrusive* solar technology, becomes an architectural material that architects are willing to use, as long as it is invisible!

This design case is the Solaris building, which is an apartment building in Seestrasse 416 Zürich, designed by Huggenbergerfries Architekten (Adrian Berger, Erika Fries, and Lukas Huggenberger are clients, architects, and electricity producers at the same time) and built in 2017.

The *mantra* of the architects was: *This is a glass building, rather than a solar building. Any photovoltaics must be invisible.*

The floor plan of the building, as shown in Figure DC-5.1, has been determined by various constraints of the context: on one side of the plot is adjacent to a main street, on the other side you'll find rail tracks. The location itself is in the vicinity of the beautiful lake of Zürich. The need to respect rules about the acoustic pollution (considering the noise of the trains and the noise of the traffic in the street), and, the intension to give each single apartment optimal solar access as well as lake views, are the premise for the "wasp waist" that characterizes the floor plan. This is the solution: the architects designed to give all apartments a view on the lake, similar solar access, and natural ventilation, avoiding problems related to the noise (Figure DC-5.2). The different exposures of the eight façades allow for a continuous solar capturing through the day and the different seasons. The chosen solution is innovative because it overcomes the standard approach of the use of the whole plot, so as to maximize the economic efficiency of the design, in favor of other sustainability and comfort factors.

In contrast to the complexity of the floor plan, the façade is quite homogeneous, as all the surfaces, roof included, are unified into a unique surface, as shown in Figure DC-5.3. This smoothens the distinctiveness of the map, and from outside the buildings look quite "ordinary." The façades have over 1,300 PV modules, which are encapsulated in glass–glass modules with digital prints on

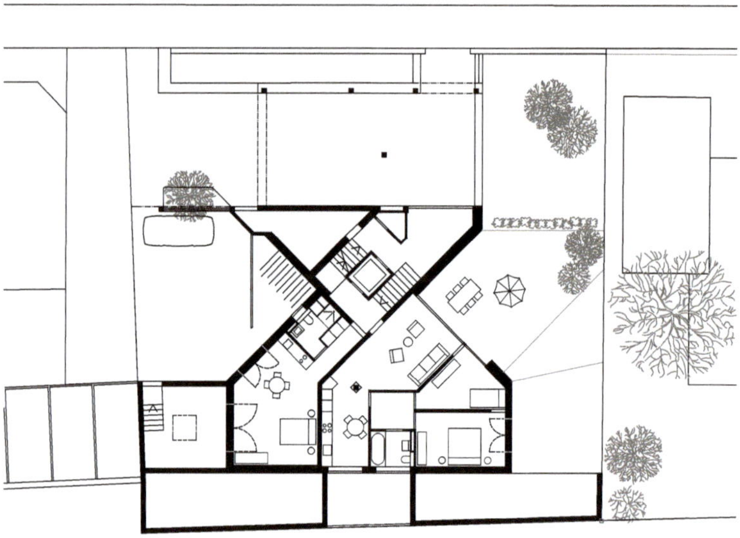

FIGURE DC-5.1 Floor plan of the Solaris building (1:200). (Courtesy of Huggenberger-fries Architekten.)

FIGURE DC-5.2 The view of the interior shows how the architects designed the floor plan, so as to give similar solar and lake view access to all the units, using the natural ventilation strategy avoiding the noise pollution. (Courtesy of Huggenbergerfries Architekten.)

FIGURE DC-5.3 View of the Solaris building from the main street. (Courtesy of Beat Bühler, photographer Zurich, courtesy of Huggenbergerfries Architekten.)

their front sides. These have an average of 10 cells, depending on the module size. A detail of the constructive solution, for using PV modules as cladding material is shown in Figure DC-5.4.

The PV modules are covered with a colored glazed surface, which seeks for a kind of glittering effect, similar to the water surface of the lake. In this way, the PV appearance is almost completely hidden, because the pattern and color of the modules are only visible under some specific angles of the incident Sun irradiation, as shown in Figure DC-5.5. The color and its density were developed over several stages by ertex solartechnik GmbH in cooperation with the university of Luzern, under the supervision of the architects. The goal was to create the optimal balance between appearance and performance of the PV system. Specifically, a high opacity of the external printed layer (that makes the PV module underneath invisible) implies a loss in energy production because the solar radiation cannot pass through the surface. Furthermore, the color itself has an effect on the power as well. Blue (the color that corresponds to the optimal solar absorption) was added to the brick red of the ceramic printing to ensure optimal power performance and creating the final brownish red color.

The choice to cover all surfaces of the building with PVs (not only the ones that are optimally oriented toward the Sun), so as to achieve uniformity of the building envelope, has some consequences in terms of generated solar energy. An estimated 14% loss in terms of efficiency of PVs (kWh/kW$_p$ generated

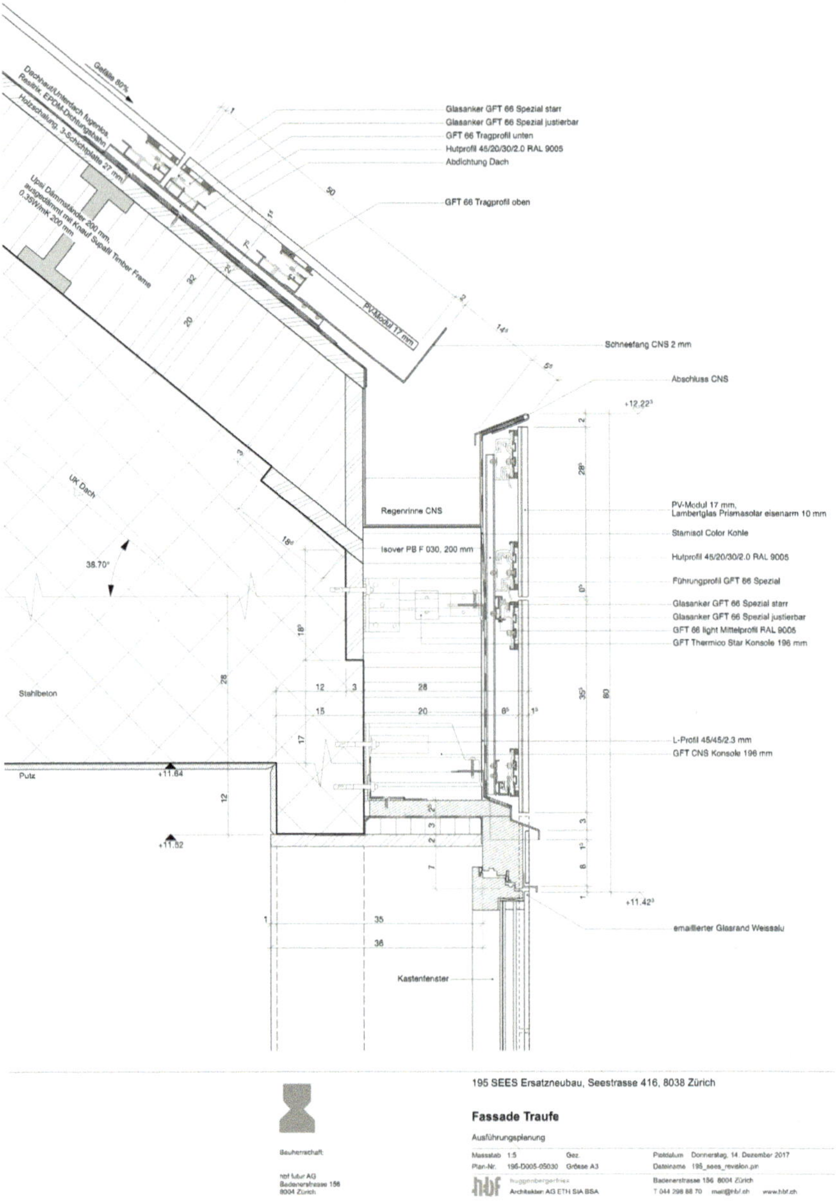

FIGURE DC-5.4 Cross section of the façade (1:5), showing how PV glasses are used as a cladding material. Modules facts: monocrystalline cells; VSG 10/5 and VSG ESG 4/5 l in different sizes; surface of modules: profiled cast glass printed with digital ceramic printing; invisibly mounted with back rails, gap between modules 6–8 mm. (Courtesy of Huggenbergerfries Architekten.)

FIGURE DC-5.5 A detail of the glittering effect, similar to the water surface of the lake, given by the use of the PV modules. (Courtesy of Beat Bühler, photographer Zurich, Courtesy of Huggenbergerfries Architekten.)

in a year), with respect to the optimal tilt angle, is the compromise the architects deliberately choose for following the architectural image they wanted to obtain.

The PV system covers about 800 m^2 (roof + façade), with a nominal power of about 72 kW_p. The glass–glass modules have digital ceramic prints with various dimensions and have been produced by ertex solartechnik GmbH in Austria.

Due to the different exposure of the modules to solar irradiance, power optimizers are required. Physically, these are small black boxes installed on the rear side of each PV module. There are 350 power optimizers, each one connecting up to 80 cells, 5.03 W_p each.

The use of power optimizers is relatively new because they are costly; however, they offer several advantages, such as:

1. Reducing the decrease of efficiency due to shading effects on the modules, which frequently occurs in BIPV with a complex morphology, or in densely populated urban areas.

   Given the spatially variable irradiance conditions on a PV module due to shading, with power optimizers it is possible to separate a complete PV module in several sections, which, from an electric point of view, function as single PV units. In this way, each single section can be operated in its maximum power point, with less energy losses in the PV module as a whole.
2. Allowing design flexibility regarding the shape and size of PV modules.

   In absence of power optimizers, all PV modules in a BIPV system should have the same number of identically sized solar cells, in order to produce the same nominal power for proper electric interconnections.

3. Allowing an easy detection of possible faults of the system, as each single unit (which is connected to a power optimizer) can be controlled separately for monitoring of its performance and efficiency.

In addition to these advantages, the connection of power optimizers in series enables the use of thinner wiring than the thick cables that would be necessary for parallel connections. This is an advantage because thick cables would have an impact of the structural and morphological aspects of a building, especially, in this case, in the thinnest part of the building (the wasp waist).

Of course, there are not only advantages; in fact, besides the beforementioned high costs, accessibility for the inspection of large numbers of optimizers can be an issue. For that reason, four main inverters have been installed in the basement of the building, each one controlling one part of the BIPV system.

The PV system has been sized for self-consumption, which is a term to quantify the simultaneity of power consumption and power generation. For instance, in this case 40% of the generated solar power is meant for immediate self-consumption. To allow this, a small battery with a capacity of 10 kWh has been installed to store the solar energy that is not immediately consumed; the remaining energy is fed into the grid. Thanks to the versatility of the PV system, mainly due to the different orientations of the arrays, the generation of the system over the day is quite constant with limited power peaks. With an energy production of about 45,000 kWh per year, this PV system completely offsets. Most of the yearly energy demand of the building is generated by the PV arrays, which are located on the roof and the south-oriented façades. Obviously, for low solar altitudes, the amount of power generated by the PV façades tends to be higher than that by the roof.

The PV system also powers an electric car that is shared by the tenants and can be booked with a Weshare app and acts as a dynamic electric storage.

5.3 CONCLUSIONS

The Solaris building demonstrates that architecture provides valid answers to our new sustainability demands toward buildings, and once again, demonstrates that buildings are unique integrators of technologies (consider PV modules, the battery and the electric cars, for instance). Moreover, new solutions are possible where the optimal solution is driven by a compromise between economical, technical, and other cultural, disciplinary requests.

If architectural solutions for sustainable and beautiful buildings are available, then other non-technological barriers, which are mainly related to the energy market, should be removed. The testimony of the architects who designed the Solaris building is that they now find PV technologies appealing enough to be used, *sexy*, but they would like to be supported by an appropriate market policy. Namely, they find that the energy, which the building generates is sold at a too-low price to the grid compared to the price for buying energy from the grid (7 €cts vs. almost eight times as much).

Actually, the best advertisement for PVs is when architects would like to propose this technology to their clients as an enabler for a zero energy bill. Unfortunately, this is not yet true in many BIPV markets around the world.

ACKNOWLEDGMENTS

The author thanks Dieter Moor and Michaela Fink (ertex solartechnik GmbH); Adrian Berger, Axel Simon (Hochparterre Verlag für Architektur, Planung und Design), and Caspar Schärer.

REFERENCES

Scognamiglio, A., C. Privato. (2008). Starting points for a new cultural vision of BIPV, *Proceedings of the 23rd European Photovoltaic Solar Energy Conference*, Valencia, Spain, 1–5 September 2008, pp. 3222–3233, ISBN 3-936338-24-8.

Scognamiglio, A., F. Garde, H.N. Røstvik. (2014). How net zero energy buildings and cities might look like? New challenges for passive design and renewables design. *Energy Procedia* 61: 1163–1166. http://dx.doi.org/10.1016/j.egypro.2014.11.1044.

6 Users' Interactions with PV-Powered Products

Georgia Apostolou

CONTENTS

6.1 INTRODUCTION

User interaction with photovoltaic (PV) products, either in a product context, such as product-integrated photovoltaic (PIPV) or PV-powered vehicles, or in the framework of buildings, such as building-integrated photovoltaic (BIPV), should be observed and analyzed, in order to better design PV-powered products. By a deeper understanding of what users' beliefs, perceptions, and thoughts are regarding these products, it can be decided what kind of developments should take place to realize better designs, functionality, and usefulness of these products. Therefore, this chapter will focus on users' interactions with PV-powered products (Apostolou 2016).[1] It will review users' expectations before they use a PV-powered product and after having used it; it will capture their experience, as well as the fulfillment of their expectations and needs.

Because of the relative newness of PV-powered products on the market, in this study we will look into the perceptions of "lead users" while not addressing all regular

[1] This chapter is based on a research study conducted by G. Apostolou from 2011 to 2016 and presented in her PhD Thesis entitled "Design Features of Product-Integrated PV: An Evaluation of Various Factors under Indoor Irradiance Conditions" (G. Apostolou 2016). The study has been published in Apostolou and Reinders (2016) and Apostolou (2016).

user groups. The reason for this choice is based on the lead-user theory of Eric von Hippel, who first introduced the term "lead user" in 1986 (von Hippel 1986). Lead users are a specific group of users: they have a need for products that are not identified yet by regular users and will become general in the market in the future. Lead users can therefore forecast future needs, propose solutions, and redesign ideas in order to contribute towards the innovation of products and services (von Hippel 1982, 1986). Logically, lead users will be the first-user category who will adapt to PV-powered products. In this chapter, lead users are observed when interacting with PV products, to identify their expectations, needs, and feelings and to finally evaluate whether PV products correspond efficiently to their wishes, abilities, and prospects.

The tested PV products (see Figure 6.1) in this chapter compromise four lighting products (the Sunnan lamp by IKEA (Inhabitat, 2019), the Waka Waka light, the

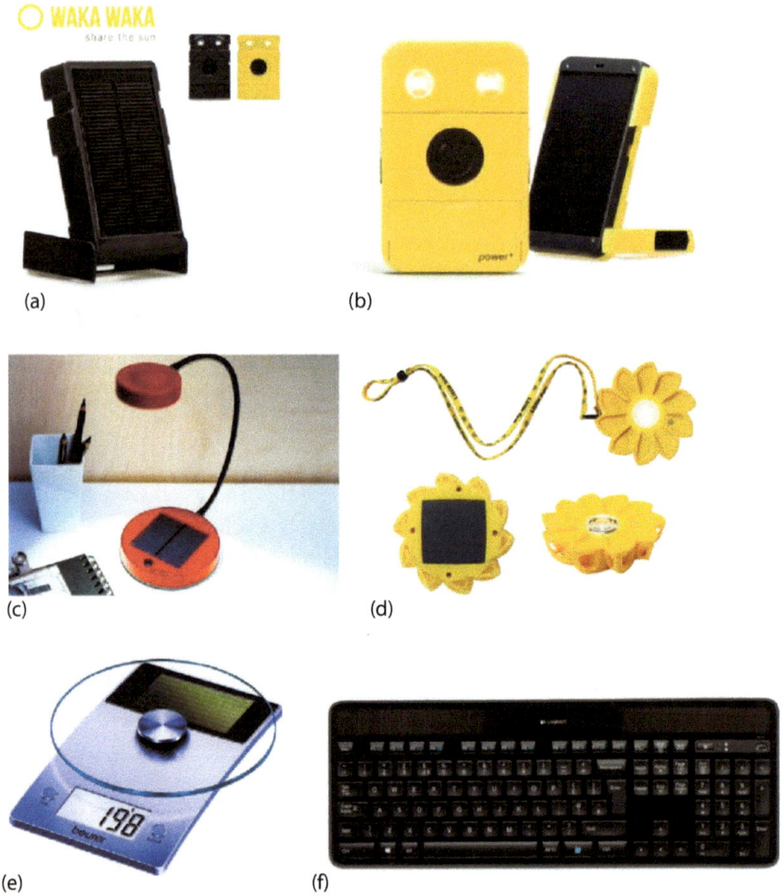

FIGURE 6.1 PV products analyzed in this study (Apostolou 2016; Apostolou and Reinders 2016): (a) Waka Waka light, ©Waka Waka; (b) Waka Waka Power light and charger, ©Waka Waka; (c) Sunnan IKEA lamp, ©Inhabitat; (d) Little Sun light; ©Little Sun; (e) Beurer kitchen weight scale, ©Beurer; and (f) Logitech solar keyboard, ©Logitech.

TABLE 6.1

Lead-Users' Reports: Interaction of Users with the Selected PV-Powered Products (2011–2014)

PV-Powered Products	Operation	Environment/ Area of Use	Reports (in Numbers)	Users (in Numbers)
Sunnan IKEA lamp	Lighting	Indoors/ Outdoors	10	50
Waka Waka light	Lighting	Outdoors	3	15
Waka Waka Power light and charger	Charging/ Lighting	Outdoors	1	4
Little Sun light	Lighting	Outdoors	3	15
Beurer kitchen weight scale	Kitchen equipment	Indoors	2	10
Logitech solar keyboard	Office equipment	Indoors	2	6
Total: 6 PV products			21	100

Waka Waka Power (Waka Waka Go Solar, 2019), and the (Little Sun, 2019)), and two PV products for indoor use (a solar kitchen weight scale by Beurer Health (2019) and the solar keyboard by Logitech, 2019). Table 6.1 gives an overview of the tested PV products and their main functions, as well as the average number of lead users that interacted with each product.

Besides the investigation of lead users' interactions with PIPV (Apostolou 2016; Apostolou and Reinders 2015, 2016), part of which is presented in this chapter, Apostolou et al. (2018a, 2018b) also investigated users' interaction with solar-powered e-bikes by conducting a field study in the Netherlands (Apostolou et al. 2018a, 2018b). The aim of that study was to capture the status of and experiences with the use of the solar-powered e-bikes. Finally, users were quite enthusiastic during their interaction with the solar e-bike and were even more interested in the solar e-bike after testing; some users even chose it as a transportation mode for their daily trips. However, despite users' interests after experiencing the solar e-bike, their likelihood to buy one was still low. The study concludes that solar-powered e-bikes have potential as a sustainable way of transportation in urban areas and cities, potentially replacing the conventional means of transport (Apostolou et al. 2018a, 2018b). In this context, one of the design cases presents a PV-powered charging station for e-bikes, which is used and tested at University of Twente in the Netherlands.

Regarding users' perceptions of BIPV, a few studies that are available (Jacksohn et al. 2019; Kurdgelashvili et al. 2019; Vasseur and Kemp 2015; Barbara van Mierlo 2017) show that several aspects influence users' perceptions regarding the adoption of PV systems by households; these are, among others, the advantages of the installation of PVs in a building context, the complexity of the construction itself, housing characteristics, the owner/user personality, possible social and demographic characteristics, and environmental concern, as well as economic factors of such an investment.

Generally speaking, capturing users' interaction with products, services, and buildings is a very demanding field of research and requires long-term observations and analysis and an adequate number of users for the field trials, which statistics usually needs to be as high as possible. Also, the control of many different variables during the tests is a challenge, in particular if aiming at acceptable outcomes and conclusions. Therefore, research studies in this field are rather limited (Apostolou 2016; Apostolou and Reinders 2016; Kakee 2008; Reich et al. 2007, 2008; Jelsma 2006; Fulton 2003). Unfortunately, after 2016, references regarding the users' perceptions and users' interactions with PIPV can be hardly found, since research activity in the field is quite narrow. As a final comment, it is therefore mentioned here that users' perceptions of PV in mobility, as well as users' perceptions of BIPV will not be further investigated in this chapter. Chapter 6 is, as such, mainly focused on user perceptions and user interactions with PIPV. In Section 6.2, further existing literature on user studies is presented. Next in Section 6.3, the methodology applied to the lead-user study of six PIPVs is presented. The results of this study are presented in Section 6.4. The chapter ends with a discussion and conclusions in Section 6.5.

6.2 LITERATURE RESEARCH ON USER STUDIES

Regarding sustainable product design, of which PIPV is a specific category, some issues of consumer behavior have been extensively discussed by several researchers (Apostolou et al. 2014, 2018a, 2018b; Apostolou and Reinders 2012, 2014, 2016, 2017; Apostolou 2014, 2016; Reinders et al. 2007, 2008, 2012b; Bakker et al. 2010; Smit et al. 2002; Vredenburg et al. 2007, 2008, 2002; Wever et al. 2008; Shackel 1984; Rodríguez and Boks 2005; Reinders and van Sark, 2012a; Kakee, 2008; Jelsma, 2006). However, there still exist some questions that need to be answered.

Jelsma and Knot (2002) pointed out that users' behaviors can be led and be determined by the design of a product (Wever et al. 2008; Jelsma and Knot 2002). This approach seems applicable to the design of PV products, since the PV products need the specific attention of the user and they could control user behavior. However, in practice it seems that users cannot easily adapt to a specific behavior, and therefore a design based on this approach could not be effective. Users prefer to be autonomous and independent regarding the use of a product, and in principle they are not willing to change their daily behavior and habits.

Rodríguez and Boks (2005) chose a rather different approach, trying to adapt the design of the product to the user's needs (Wever et al. 2008; Rodríguez and Boks 2005). Regarding the design of a PV product, it seems that the approach by Rodríguez and Boks (2005) is more appropriate than that by Jelsma and Knot. It could be more fruitful to design a PV product, according to the user's needs and expectations. However, in order to achieve this, it is important to discover and analyze what users expect from these products. Furthermore, Vredenburg et al. (2002) pointed out the quality of products by focusing on users' abilities and wishes, aiming to enhance user interaction with the product.

Rooden and Kanis (2000) suggested an advanced model for user-product interaction, which is based on the model by Shackel (1984), and contains user's features, product's features, and details about the relation between the user, the product, and

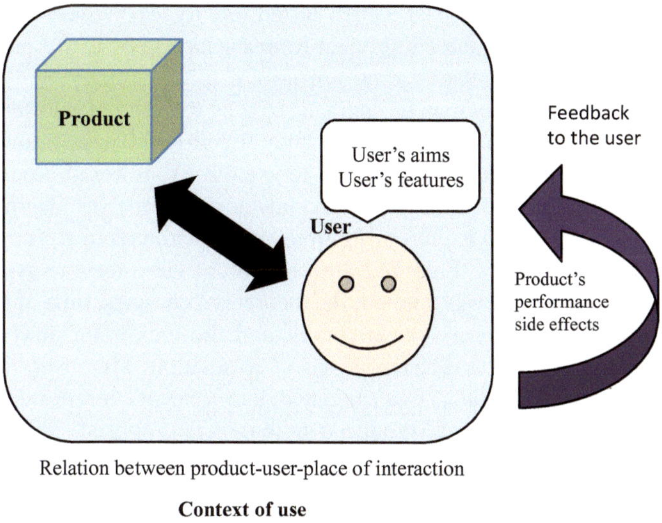

FIGURE 6.2 The model of Rooden and Kanis (2000) illustrating the user-product interaction (scheme designed by the author). (From Wever, R. et al., *Int. J. Sust. Eng.*, 1, 9–20, 2008; Rooden, T. and Kanis, H., Anticipation of usability problems by practitioners, *Proceedings of the Human Factors and Ergonomics Society 44th Annual Meeting*, San Diego, CA, 2000.)

the place of interaction (Wever et al. 2008; Shackel 1984; Rooden and Kanis 2000). The model also offers feedback information to the user on the performance of the product, as well as the possible side effects. This model is demonstrated in Figure 6.2.

The model of Rooden and Kanis (2000) will be used in our study on the lead-users' interactions with PV products.

Sanders (2002) talks about the different experience characteristics of users and divides them in four categories: the "obvious," which depicts things that people say or think; the "visible," which refers to what people do; the "unspoken" or "tacit," which are things that people feel or are aware of; and the "dormant," which are people's dreams (Sanders 2002, 2006a, 2006b). In this study, Apostolou (2016) attempts to uncover all the above aspects and focus mainly on the unspoken and dormant features, which are important for designers, through lead-users' interactions with the PV products, their daily notes on the workbook, and their thoughts about the products they used.

6.3 METHODOLOGY

The study on the lead-users' interactions with PV products took place during the academic years 2013–2014 and 2014–2015 during a PV workshop, at Delft University of Technology, in the Netherlands. The sample of the respondents consisted of 100 lead users (75 men and 25 women) at the age of 20 to 35 years old. Approximately 90% of the respondents were Dutch, and the remaining 10% originated from the EU or India. All the participants were bachelor students of the Studies of Industrial Design

Engineering. Before the field trial, which lasted one to two weeks, the lead users could choose and decide which PV product from a selection of 12 PV products they would like to use.

Respondents of this study may be characterized by the term "lead users" because it was defined in the Introduction of this study. Lead users were asked to follow some specific tasks with the products; first to use the PV-powered product in their daily routine and then to disassemble and analyze it, in order to identify its main components and to evaluate its feasibility from a technical, practical, economic, and environmental point of view. From a technical point of view, the energy-conversion efficiency of the solar cells was defined, the theoretical charging time of the battery, the current-voltage and power-voltage curves, and the maximum power point, by conducting measurements at different levels of irradiation. Moreover, the application of solar cells in the products was evaluated, and concepts or redesign ideas for a potential improvement of the PV product's system were proposed. The respondents used the PV product as a part of their daily routine. The issues that they had to note during the trial week were the following:

- Their initial expectations from the product before using it
- Evaluation of the product after using it for a short period of time
- User pattern and using frequency
- The ease of usage and general functioning of the product
- Emerging frustrations or feelings of satisfaction/dissatisfaction with the PV product
- Suggestions for improvement
- Ways to use the product in their daily life

The methods that were used to gather the research elements regarding the respondents' comments before, during, and after the field trial, as well as their frustrations, problems, and suggestions, were notebooks for writing down their daily routine and their feedback during their interaction with the product. These notebooks were presented in formal reports written by the students and were used by the authors of this research. The authors organized the data collected from the reports, categorized them, and drew conclusions. Furthermore, lead-users' interactions were based on self-observation during their daily routine with the product, while a direct observation method was applied during the analysis of the technical features of the PV products, as well as during the performance tests.

Besides the preceding issues, a questionnaire of 17 questions was set up in order to collect information about lead users and their interactions with the PV products. Two sets of questions were formed: the first one consisted of closed questions, where the possible answers were set on a scale of satisfaction (from 0–low to 5–high); whereas the second set consisted of open questions, where the respondents were asked to elaborate on their opinions and thoughts.

The questionnaire was first distributed to the lead users, since its aim was to outline their first reactions and thoughts about the PV products after the respondents' early interactions with the product. Lead users first answered the closed questions of the questionnaire, and after a few days of living with the product, they continued with the open questions. After the questionnaire was completed and during the field trial, the lead users were self- or directly observed interacting with the product, and by the end of this period they already had prepared the report with their actual experience. Lead users' answers from the questionnaire will be presented in the form of statistics. Further results of this study, such as possible redesign of PV products or best-fitted user context with a specific product, as proposed by the respondents in their formal reports, will be analyzed and discussed in the results section.

6.4 RESULTS

6.4.1 Analyzing Lead Users' Answers from the Questionnaire

The results depicted the lead users' expectations at a time before using the product and their evaluation after the field trial. It is interesting to observe the difference between the two stages. During the first contact of the lead user with the product, and before the field trial start, lead users criticize the outlook of the product (e.g., the design, color, materials, size) and they try to predict its function and usefulness. Approximately 60% of the respondents feel comfortable with the product and consider it as "a nice gadget" to use. First impression is positive. However, there are doubts concerning the functionality and performance of the product. Performance is defined as the level in which the intended function of the product performs well (e.g., typing using the keyboard). The product's performance does not refer to the performance of the product's PV cells.

After the field trial, lead-users' feedback in the form of written reports and answered questionnaires mainly concerns the product's performance; approximately 40% of the respondents are totally unsatisfied (see Figure 6.3), 38% find the product totally useless (see Figure 6.4), 60% find the design of the product of bad/low quality

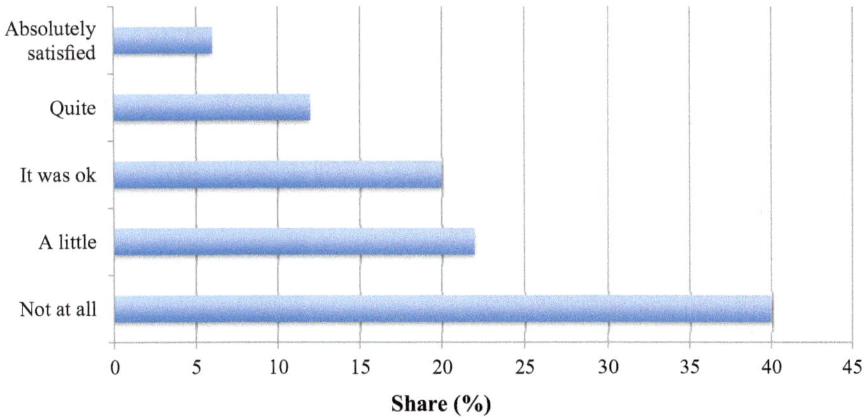

FIGURE 6.3 Lead-users' answers (share %) to the question: Are you satisfied by the PV product's performance? Number of respondents $n = 100$.

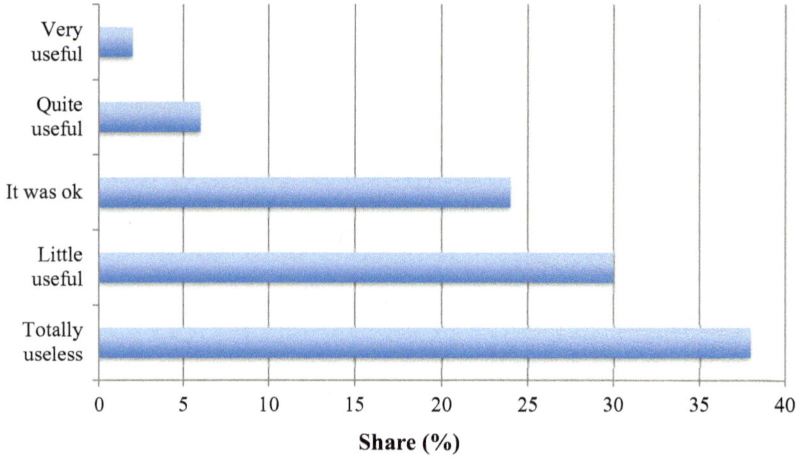

FIGURE 6.4 Lead-users' answers (share %) to the question: How useful did you find the specific PV product? Number of respondents $n = 100$.

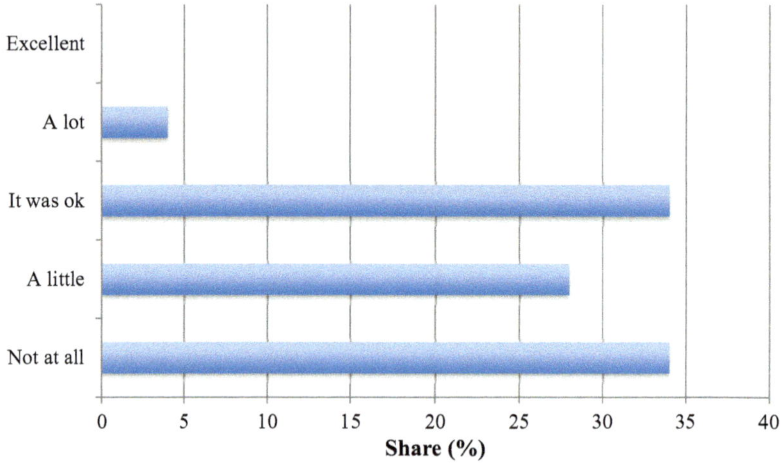

FIGURE 6.5 Lead-users' answers (share %) to the question: Did you like the design (look) of the PV product you used? Number of respondents $n = 100$.

(see Figure 6.5), 54% believe the design of the product is quite simple and can easily be used by everybody (see Figure 6.6), while only 4% find it difficult to use the specific product (see Figure 6.7). Approximately 88% of the respondents would not buy the PV product or propose it to a friend (see Figure 6.8), and 70% believe that the price of the PV product does not correspond to its quality and performance (see Figure 6.9). Table 6.2 presents some comments regarding the general evaluation of the PV products, based on the lead-users' remarks (Apostolou 2016; Apostolou

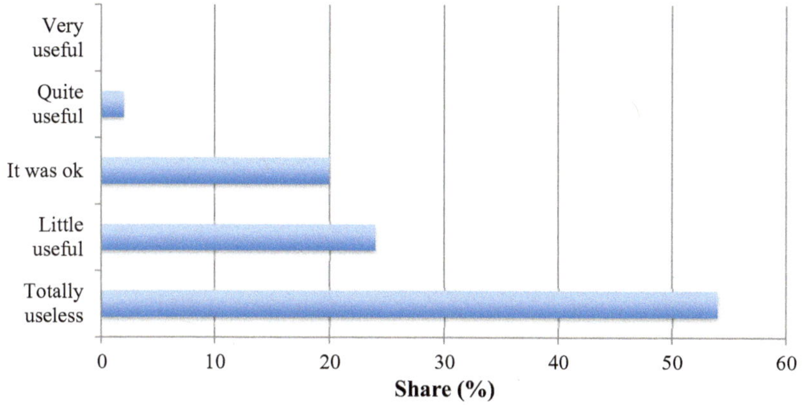

FIGURE 6.6 Lead-users' answers (share %) to the question: Did you find the product's design complex? Number of respondents $n = 100$.

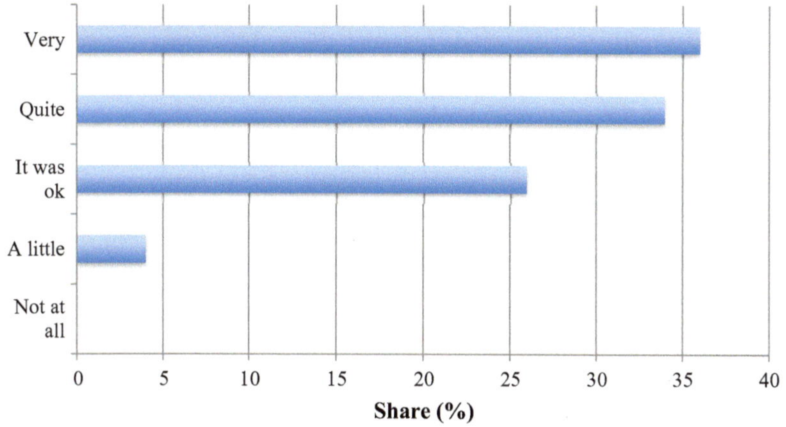

FIGURE 6.7 Lead-users' answers (share %) to the question: How easy was it to use the specific PV product? Number of respondents $n = 100$.

and Reinders 2016). Last but not least, approximately 66% of the respondents would prefer a product that can be charged by a cable with a plug, instead of a PV-powered product (see Figure 6.10).

Main results show that the lead users need more reliable PV products, made with materials of good quality that have an interesting and appealing design and perform sufficiently. Instructions on the packaging/casing of the product and reliable expectations from the manufacturer/designer seem to be important to lead users, in order

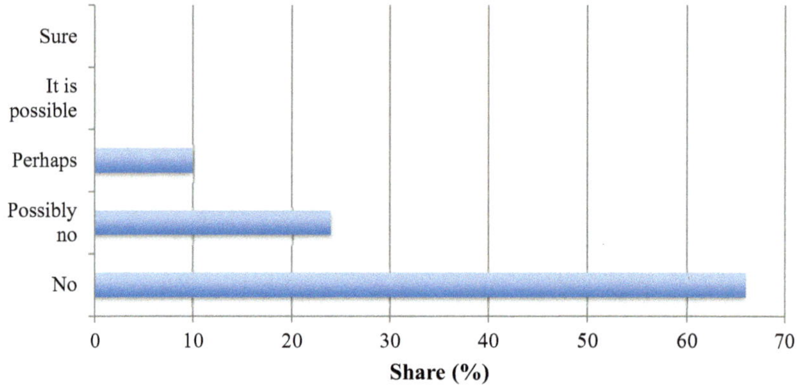

FIGURE 6.8 Lead-users' answers (share %) to the question: Would you buy the PV product you used during the field trial? Number of respondents $n = 100$.

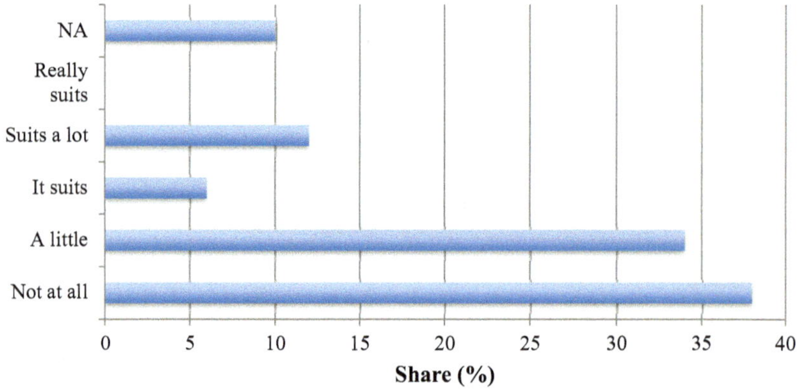

FIGURE 6.9 Lead-users' answers (share %) to the question: Do you think the price of the product corresponds to its quality? Number of respondents $n = 100$.

to know what to expect from the product and how to use it. Respondents are willing to pay money and buy a PV product if it is useful and works properly. Furthermore, it is noticeable that lead users are quite positive with PV products that have an environmentally friendly or a social character (e.g., donations to the developing countries when buying a solar-powered lighting product, such as the Waka Waka or the Little Sun light).

TABLE 6.2
Example of the General Evaluation of the PV Products, According to the Perception of a Single User's Feedback

	Sunnan IKEA Lamp	Waka Waka Light	Waka Waka Power Light and Charger	Little Sun Light	Beurer Kitchen Weight Scale	Logitech Solar Keyboard
Design	Simple, Practical, unattractive	Strong/Simple	Strong/Simple	Interesting shape, attractive	Simple, not efficient	Premium, high quality
Usefulness	Cordless, short battery lifetime	Cordless, necessary to developing countries	Cordless, necessary to developing countries	Cordless, necessary to developing countries	Cordless, nice gadget	Cordless, power independent
Performance	Insufficient, low lighting levels	Sufficient, strong light	Rather sufficient	Depends on the charging	Not precise	Sufficient enough, battery status 100% full
Sales Price	Affordable	High	High	High	High	High, but worth the money
Portability	+	+	+	+	+	+

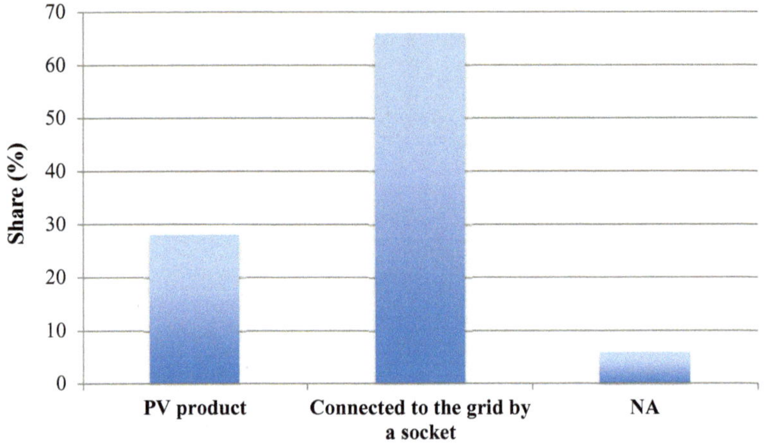

FIGURE 6.10 Lead-users' answers (share %) to the question: What would you choose to buy, a PV-powered product or a product with a cable and plug? Number of respondents $n = 100$.

6.4.2 Lead-Users' Feedback

We asked for respondents' personal opinions. Below you will find a selection of statements that indicate lead-users' personal opinions after their interaction with the product; these are stated in the students' reports (Apostolou 2016; Apostolou and Reinders 2016).

"A multi-function PV product is more desirable (e.g., lighting and charging function)."

"I would prefer to buy a product without PV cells. The PV cells must be removed and placed in the sun every day (outside). If I could just leave it inside, then I would reconsider it."

"I would buy the PV product. I do not like wires, and batteries are always nowhere when needed."

"I think the specific PV-powered light is a useless product in Holland, due to the lack of sunlight, but in Africa it would be a great product."

"I find it a bit frustrating that the product works well only with heavy sunlight."

"I would definitely buy the PV-powered product. I am happy to spend a little more money for eco-friendly products that use renewable energy."

"I liked the design and the idea of using a PV-powered product. What I did not like was the fact that it did not work late at night, when the sun was down."

"I would buy the grid-connected version of the product, because it is less expensive and it performs better than the PV-powered product."

"The design of the product is ugly, and the battery pack did not charge. I am really disappointed by its performance. It is the worst product of the branch."

"What wouldn't I change on the PV product…design, battery capacity, materials, color, use…"

"The overall product needs improvement, but the concept is good. A redevelopment could deliver a better product."

"I would buy the PV product just for fun, if it was cheap more than wanting to use it,"

"Grid connection is more reliable than PV."

It can be concluded from the above statements, reflecting only 20 % of the total, that they are very diverse and represent both negative and positive opinions.

6.4.3 Analyzing Lead-Users' Interactions with the Tested PV Products

In this section, the six tested PV products (see Figure 6.1) are separately presented and the lead-users' interactions with them is discussed. Respondents' thoughts and ideas for redesign are also addressed, as they were evaluated in the students' reports.

6.4.3.1 IKEA Sunnan Lamp

a. Product's features

   The Sunnan lamp is a wireless product that is portable and quite flexible to the user. There is only one button present on the base of the lamp, which makes the lamp's operation quite easy. Furthermore, a movable steel arm is present, as well as three LEDs. The solar cell of the lamp can be detached from the lamp and be charged outdoors or indoors near the windowsill. The Sunnan lamp can be used as:

   - a desk light; for reading, writing
   - a garden table light; using it outdoors on the garden table, when natural light is insufficient
   - a bed light; using it on the bed, while reading a book
   - a flashlight; due to the product's portability, it can be used as a torch. However, the shape is odd and it is rather difficult to hold the lamp

b. Lead-users' expectations before use

   Initially, the fact that there are no wires permits lead users to place the lamp where they want, without keeping it near a power socket. Second, it seems easy and convenient to use the lamp, since there is only one button to turn it on/off. Third, the movable steel arm enables the lead user to adjust the angle of the incoming light. Finally, the lamp uses three LEDs, which should be more than enough for a reading light. The solar cell is removable, and the user needs to place it under sunlight, instead of doing so with the whole lamp.

c. Lead-users' experience and feedback during use

   After the field trial, the lead users reconsidered some of their initial expectations. Comparing the light to a normal light bulb, one issue immediately arises: the light does not diffuse. The light of the LED only covers a circular area of about 50 cm (when aimed right down from the highest point), and

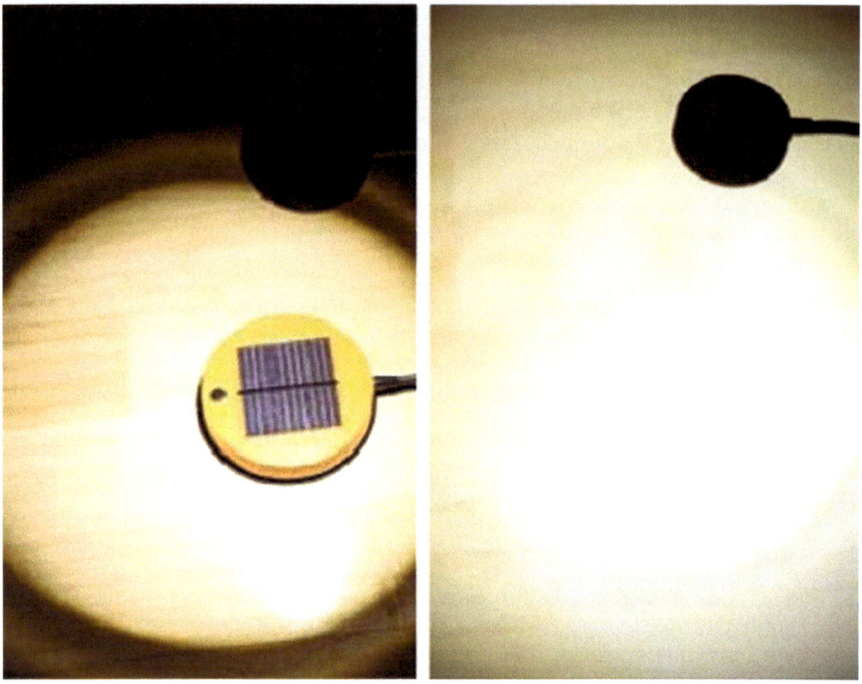

(a) (b)

FIGURE 6.11 (a) Current situation vs. (b) ideal situation. In the ideal situation, the light diffuses more, which results in a more gradual transition between light and dark.

beyond these perimeters the surface is totally dark. The light is concentrated on one spot and does not diffuse in any direction. According to the respondents, this is a drawback of the product (Figure 6.11).

d. General conclusions and discussion after use

The lamp is solar powered; therefore, the lead user needs to take into account that the battery is charged sufficiently to use it 2 hours a day, three times a week for desk activities and half an hour a day, five times a week and more, when necessary. Following the manual, the lamp needs to be charged for minimal 9 hours in sunny conditions and 12 hours in cloudy conditions to function for 3 hours. On the shortest day of the year, there are less than 7 light hours, which prevents full charging of the product. For that reason, the lamp cannot be used to full potential.

The field trial determined that the actual burning time of the lamp is much longer than is guaranteed, so it is expected that this will compensate the shorter period of exposure to daylight.

It is anticipated that with 2/3 of the charging time the duration of light burning will drop with 2/3 also, to around a burning time of 5 hours.

Nevertheless, after the burning time of the 3 hours mentioned in the manual, the lamp color gets a different, but warmer tone. This tone is even more comfortable than the clinical light color. Therefore, it can be stated that the lamp supplies the demanded usage from the target groups, when charged outside.

In reality, lead users will charge the batteries indoors. The irradiation will drop dramatically by a factor of 20, which means those 5 hours of burning time anticipated would not be reached by charging it in one day. It is neither expected that indoor charging can reach the burning time for the daily use of maximum two and a half hours desired. It will be even more difficult to satisfy the needs for unexpected higher use.

Overall, the resilience of the light was collectively observed. It would not easily run out of battery power. A significant difference between the high- and low-power performance state was noticed by the time respondents learned how to adequately power-up the product. The drop-off in power was so noticeable that lead users were forced to reconsider how useful the light was, once the brightness began to dim as the batteries discharged. This is illustrated in the contrasting images in Figure 6.12.

To conclude, it is favorable to view the Sunnan lamp as two different products combined: a desk lamp and a social product. When considering it as a desk lamp, it can be said that it fulfills the expectations and adheres to the performance that was indicated in the manual. On the other hand, as a desk lamp it is slightly unwieldy because it requires daily charging, and hence its usability depends largely on the respondent's discipline (in putting the solar panel outside) and the grace of the weather. As such, its shortcoming is that it is slightly unreliable, and compared to other lamps, it contains limited user interaction (it only has one button, while a lot of desk lamps have multiple brightness settings).

As a social product, it is slightly more successful, though it is not completely obvious to the consumer. The policy of IKEA (the Light in the Dark project) stipulated that for each Sunnan bought, one is donated, so that children in countries without electricity can still read/see after sunset, with the use of the Sunnan.

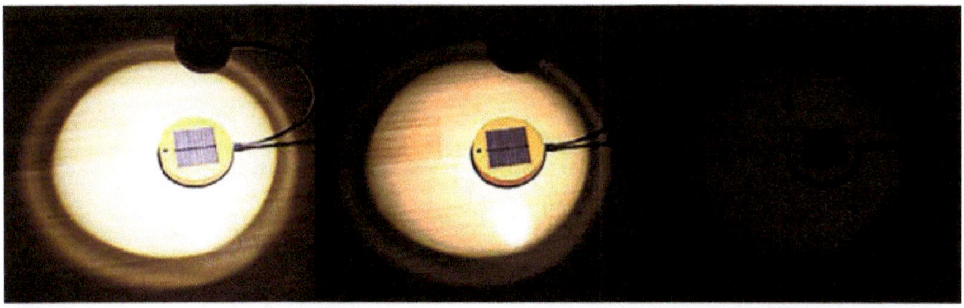

(a) (b) (c)

FIGURE 6.12 Illustration of the performance of the Sunnan solar-powered lamp after fully charged and after several hours of using it: (a) bright, (b) dimmed, and (c) off mode of the Sunnan lamp.

This is a noble act, and combined with the educational power of the Sunnan in bringing awareness of solar energy and the power of LEDs, this lamp has definitely succeeded as a social product. However, there are still some improvements possible when it comes to making the consumer aware of this side of the Sunnan desk lamp.

e. Redesign IKEA Sunnan lamp, as proposed by 50 lead users
In terms of general improvements for the Sunnan lamp, lead users suggest: a power toggle switch (Hi/Med/Low), so that different values of brightness could be selected when using the lamp according to the type of use. This would improve the length of use, when full brightness is not necessary. A folding stand like on a picture frame at the back of the battery pack could give it a better position toward the sun while charging. Easy access to the batteries is also essential. Another improvement could be on the width of the light beam and the light intensity (Figure 6.13). The beam should be significantly improved, because the current one is insufficient. Moreover, an LED indicator to display the battery level, or the charging status, would give the lead user important feedback about repositioning the product or reducing the power for preservation. For indoor use, the respondents believe that the product could be easily placed near a window, by using a suction cup, and it might be more efficient to select a new type of PV cell that could achieve better performance indoors.

FIGURE 6.13 Some suggestions for improvement of the Sunnan lamp, according to lead users.

6.4.3.2 Waka Waka Light and Waka Waka Power

a. Products' features

The Waka Waka light is a small electric light charged by a solar panel on its back surface. Two small LED lights can be used in three light intensity settings. There is only one button on the product; the user can push it multiple times for three different light intensities. Pressing the button for three seconds makes the product give off a Morse code SOS signal.

The product is aimed at people who live off the grid, although it can be used for different uses, such as camping. The product also has a stand, which can be used either to support the product on its own, or to prop it on a bottle by using the hole on the stand.

The next version of the Waka Waka light is the Waka Waka Power, which is a strong and solid solar charger, able to charge almost all (smart) phones or other small electronic devices within a few hours and to offer around 150 hours of lighting.

The Waka Waka Power has two target groups:

a. first-world and
b. third-world country people

The first group could also be divided into two subgroups:

i. people who buy the Waka Waka as an act of charity and
ii. people who buy the Waka Waka, because they actually need a solar charger and for whom the charity is an emerging subsequent

For the last group of people, it is likely to assume that they bought the Waka Waka because they need a light and portable charger during outdoor activities, such as camping, where they lack the possibility of charging their phones or other gadgets.

b. Lead users' expectations before use

The initial lead-users' impressions of the product's appearance was rather strange. The shape of the product looks odd and does not enhance the functionality of the product. However, the Waka Waka looks rugged, giving it a durable appearance. It is expected that the product can be used in two main scenarios. In the first scenario, the product will be used daily by people with no access to other sources of electricity (Figure 6.14). The product can then be used either to bring light to an entire room, or for specific everyday activities that require light, for example, reading. In the second scenario, the Waka Waka will be used by people during camping or other similar situations, where they lack easy access to electricity (Figure 6.15).

c. Lead-users' experiences and feedback during use

First, the Waka Waka was used by the lead users of the field trial as a nightstand lamp. The emitted light was quite bright, with a cold, blue tone, and it was uncomfortable when reading or looking straight at it. Moreover, the product requires no indication regarding the status of the battery. It is easy to understand that the product is not originally intended for the context that was used it in and would be more useful in different scenarios. When there is access to electricity, it seems that the product is less useful.

FIGURE 6.14 One of the multiple uses of Waka Waka light is for reading in developing countries. The product can also be placed on the top of a plastic bottle, ©Waka Waka.

FIGURE 6.15 Waka Waka Power placed on a plastic bottle of water, ©Waka Waka.

d. General conclusions and discussion after use
The Waka Waka is able to perform its functions (lighting and/or charging). However, regarding indoor activities, an alternative, non-solar-powered product could also be effective. For outdoor activities and third-world countries, the Waka Waka Power is quite an essential product. Designers thought of sustainable solution beyond PV cells but also used recycled plastics.
Furthermore, it is a very good initiative to offer Waka Waka to people in countries that really need light in the dark.
Obviously, the people in first-world countries pay indirectly for this charity project, but it is still an affordable product compared with competitors. Furthermore, it is a benefit that the Waka Waka is produced locally, when possible, so there is less transportation for shipping the products to the development countries. According to the respondents, there is still room for improvements for the Waka Waka (e.g., quality of light, brightness of the lamp, positioning of LEDs, battery status indicator, better positioning of the product).
e. Redesign the Waka Waka light, as proposed by 15 lead users
From a practical point of view, the design of the Waka Waka light is quite minimalistic. It is an easy-to-use product, which works intuitively albeit with room for improvement. An option for upgrading would be to let the two LEDs be able to separately be pointed at any direction, no matter the position of the housing.
The well-known USB snake light could serve as an example, although this technique would renounce the protection of the LEDs and make it more vulnerable. Redesigning the LED placing could be in the form of two spherical ball hinges, which enables both LEDs separately rotation and gives the user the option to aim the lights more specifically. This means that the Waka Waka light can still stand on a surface, while the separate LEDs can easily be pointed at any direction in order to illuminate a larger area. The big drawback of this adjustment is that it would probably cost a lot more money to develop and produce it, due to the additional hinges, in which the LEDs have to be integrated. One of the discomforts the light gives to its users is the bright peripheral light, meaning the light that directly reaches the eye if it is not placed behind the user. Placing a cap around the lights could both decrease the annoying peripheral light and also increase the amount of light, where it is actually needed. Thus, a suggestion for the redesign would be to make a small cap around the lights, so it does no longer emit light both upward and sideward. Placing the lights a little deeper in the product and making a cap with the shell of the product could achieve this. This could also be a benefit for the LEDs because these will be better protected.
f. Redesign the Waka Waka Power, as proposed by four lead users
The redesign of the Waka Waka Power could contain a solar panel with a bigger surface (Figure 6.16), and possibly a new battery (1.5 times bigger than the current). These updates are necessary for the product to work properly and to support the charging of the new generation of (smart) phones.

FIGURE 6.16 One of the proposed redesigns of the Waka Waka lamp, according to lead-users suggestions; the solar-panel area is extended.

6.4.3.3 Little Sun Light

a. Product's features

Olafur Eliasson designs the Little Sun, a small and independent source of light that can be used anywhere.

When charged during the day, the light can be used during the evening or night. The product is intended for people with no access to electricity. To make it affordable for the target group, the light is sold in Western countries, the revenue of which is used to reduce the prices in off-grid communities. The Little Sun is a product targeted at third-world communities without electricity, which means they do not have access to electrical lights in the evening. Light is necessary for working, studying, or even just being together. A wood fire or kerosene lamps are usually used instead of electricity in third-world communities. Both these light sources are dangerous, not only because they emit toxic gases but also due to their fire hazard. This is where the Little Sun tries to help, allowing people to have a durable, safe, and easy-to-use light source.

b. Lead-users' experiences and feedback during and after use

Little Sun emits strong light, which can be useful in multiple situations, and it is capable of lighting up an entire room. However, it is uncomfortable to handle because it is a bulky product with sharp edges. The on/off switch can be hardly found in the dark. The chord that the product contains can be used to hang it into a hook (see Figure 6.17a, b).

FIGURE 6.17 (a) Little Sun light is turned on; (b) Little Sun PV-powered light is placed near the windowsill for charging; and (c) disassembling the PV product during the field trial. All parts of the product are visible, including plastic components (casing), cables, screws, and PV cell.

The brightness of the Little Sun decreases from 100% to 0% after approximately 7 hours of use, as presented in Figure 6.18. Generally, 7 hours of use is considered a lot, and therefore the performance of the product seems to be satisfactory. However, the intensity of the light is not stable during the 7 hours of use. After 2–3 hours, the intensity becomes less strong and the light is dimmer.

c. Redesign the Little Sun light, as proposed by 15 lead users

In order to redesign the Little Sun, first it is necessary to abstract the problems. One of the main issues that users have is the safety while using the light. The sunflower shape is very decorative but has sharp edges. This is not ideal for little children because it is possible that they could hurt themselves while trying to use the lamp in the dark. Therefore, a new design of the product is proposed with more curved surfaces (see Figure 6.19). Furthermore, a charge indicator with an LED light could be added in the product, such as the users could receive feedback regarding the status of the battery. A stand for better

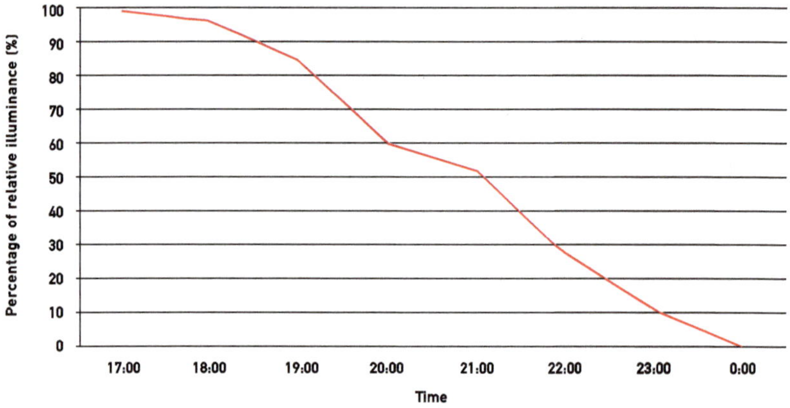

FIGURE 6.18 The percentage or relative illuminance of the Little Sun goes down from 100% to 0% after approximately 7 hours of use.

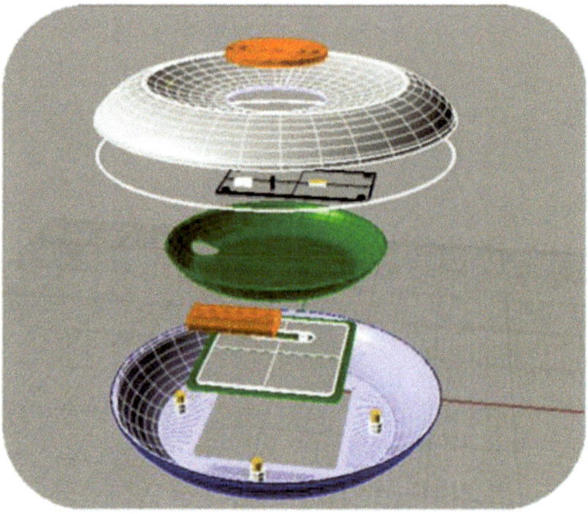

FIGURE 6.19 A redesign of the Little Sun, according to lead-users' suggestions.

positioning of the product is also essential. Last but not least, lead users propose the addition of a USB port in the product, which could offer the possibility of connection with other devices for extra charging.

A brand-new design of the product is shown in Figure 6.19, according to the lead-users' suggestions for improvement.

6.4.3.4 Beurer Kitchen Weight Scale

a. Lead-users' expectations before use
Purely from the first appearances, the product seeks to satisfy the ideo-pleasures of the respondents, fulfilling aspirations to feel "eco," both through a purchase of the product, as well as owning the product. This is achieved through the idea of a solar-powered scale, presenting itself as an "eco" and

sustainable alternative. The user will gain a greater appreciation for the product, increasing relationships and facilitating the ability to strike up a conversation about the product, also because it looks and appears modern.

b. Lead-users' experiences and feedback during use

According to lead users, it seems that the scale works fine and precisely during the day with sunlight, and it does not need any charging. It works each time you need it, just as the producers promise. The scale weighs small amounts up to 5.5 kg and has a graduation of 0.1 g (Figure 6.20). On the display, a battery bar is showed on the left top, indicating that the battery stays constantly half full even after 4 hours of sunbathing. When using the scale during the evening, the product faces difficulties catching enough light.

c. General conclusions and discussion after use

The power on time depends on the light intensity, whereas the discharge time does not. This suggests that there may be a capacitor inside the product. The purpose of the capacitor is to power the scale, when the stream of light hitting the panel is disrupted. Being powered by a capacitor may be an issue in the scenario that a large object is placed on the scale, covering the panel and disrupting the direct light. This is not a wished feature, since it gives a limited amount of time to weigh the object. The kitchen scale weighs with an acceptable order of accuracy for its use purpose. However, this uncertainty may increase, when the object's center of mass is not placed in the middle of the scale, since it has only one sensor in the middle. If the center of mass is placed off-center, it creates a moment, increasing the measured mass. This is not an issue for the scale's functionality, since while cooking or baking, one gram more or less won't make any difference.

Overall, the scale does not meet its initial expectations; at least half a minute is required before it can be used. In the case that no natural light is available, an alternative light source is required. The fact that a more-consuming energy source is required to power the product is not as efficient as directly taking the required power from the grid.

FIGURE 6.20 The Beurer kitchen weight scale is being used during the field trial.

d. Redesign the Beurer kitchen scale, as proposed by 10 lead users
The most-efficient way of improving the product's performance, without drastically altering the required technology, was to allow the PV cells to catch more light. The first way to achieve this is by removing the "hovering" glass plate. In order to avoid the shadow of the object that is being weighted to block out light for the PV cells, the PV cells were distributed over the surface of the product in a different way, as Figure 6.21 illustrates. In that way, they cannot both be fully blocked out, which means that there should

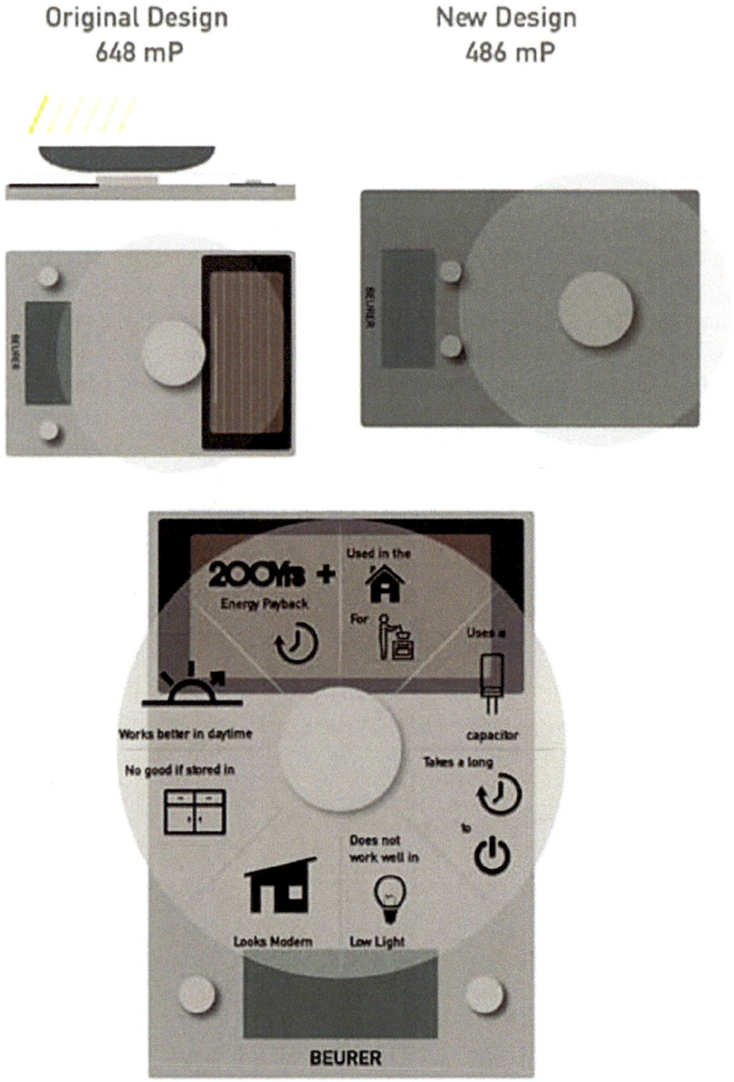

FIGURE 6.21 Redesign of the Beurer kitchen weight scale, according to lead-users' suggestions for improvement of the initial design of the product.

always be some power available. The new design that is proposed would preferably have a quite similar appearance to the old one, which is a quite smooth and modern look.

6.4.3.5 Logitech Solar Keyboard

a. Lead-users' expectations before use
Any computer or tablet can use Logitech solar keyboard. It performs absolutely powered by the sun or indoor lighting, and there is no need for charging. It is possible to be used with all kinds of digital devices. It looks very sleek and professional, and it is expected that the intended use is in a work environment (Figure 6.22). Due to its portability and the solar powering, it does not need to be added to the power grid, and it can be used everywhere. The PV cells make the product absolutely independent from the grid and to eliminate the battery changing. Logitech claims that the keyboard has a three-month battery life with no light. Taking into account that a keyboard does not take much power to operate, it will most likely not drain the battery too fast.

b. Lead-users' experiences and feedback during use
Generally, the keyboard worked very well, and the respondents were satisfied with its performance. They felt that it is a very beautiful design, both in look as in user friendliness. Furthermore, the fact that it is solar powered is an added bonus. During the testing, the keyboard had a light interaction of less than 1 hour each day. At the fourth day of the testing, the solar application (see Figure 6.23) showed for the first time a different value than 100%—a 99% state of charge (SOC) of the battery. This means that with a light use of the keyboard, the battery would be able to run for one full year without charging or in total darkness. However, there are more active users who use the keyboard more hours per day, but even in that case, the battery would still be able to discharge at around 1% to 2% each day, with then a total operation time of approximately 3 months. However, the cases that are described here are quite extreme, since the PV keyboard would normally always recharge the battery by using the installed solar cells. After the test, the battery showed the next day an SOC of 100%.

c. General conclusions and discussion after use
The keyboard worked perfectly during the period when it was used. The PV panels integrated in the product did their work sufficiently, by keeping the battery charged up at 100% almost continually. The simulations of

FIGURE 6.22 The Logitech K750 PV keyboard, ©Logitech.

(a) (b)

FIGURE 6.23 The solar application of the Logitech solar keyboard, showing the light brightness and the charging status of the battery (a) before starting the test, and (b) after 4 days of use.

the power management of the PV keyboard showed that even by a more intensive usage of the PV keyboard, the configuration of battery, and PV cell by the manufacturer would in the worst case ensure at least 13 days of use till the battery would run out, under low lighting conditions. The combination of battery and PV cells made the product totally power independent from the usual keyboard, which are bound to a periodic change of batteries. The negative criticism toward the PV keyboard was insignificant and subjective regarding small issues about the product's design.

According to lead-users' opinions, the PV keyboard was praised about its design, but finally it was also criticized about its price. Most respondents would not wish to pay double price for a product, just because of the integrated PV cells. However, even though the product is quite expensive, it seems reliable. The support and the service, which one can get from Logitech, is also valuable. The manufacturer offers a 3-year warranty and also a very good forum and website where every user can easily share with others their problems and experiences.

The solar application is also very useful since it keeps the user always informed about the battery's SOC and also about the indoor illumination.

d. Redesign the solar keyboard by Logitech, as proposed by six lead users
Generally, most respondents were satisfied with the keyboard's design. However, the easy transportation of the product is a feature that could be improved. Therefore, to make it easier to transport the keyboard, the removal of the number pad is a possible choice. This makes the keyboard much shorter and easier to transport, as the main focus now lies in the use in combination with a tablet computer. This means that the USB connection could be replaced by a connection via Bluetooth. Because the Bluetooth connection uses more power, and given the fact that the keyboard was overpowered, but reliable during the field trial, it seems that the system could support the Bluetooth connection. In a marketing sense, this small adjustment will make the keyboard more versatile because it will be introduced to the tablet market that mostly uses a Bluetooth connection. Table 6.3 is a comparison of the tested PV-powered products, according to 100 users' feedback.

The difference in quality per feature is arranged as good (+), medium (+/–), and bad (–) in Table 6.3.

TABLE 6.3
Product Comparison, According to 100 Users' Feedback

PV-Powered Product/Number of Users per Product *n*	Sunnan IKEA Lamp *n* = 50	Waka Waka Light *n* = 15	Waka Waka Power Light and Charger *n* = 4	Little Sun Light *n* = 15	Beurer Kitchen Weight Scale *n* = 10	Logitech Solar Keyboard *n* = 6
Form	+/–	+	+	+	+/–	+
Compactness	+/–	+	+	+	–	+
Use and repair	+	+	+	+/–	+/–	+
Safety	+	+	+	+	+	+
Solidity	–	+	+	+/–	–	+-
Price affordable	+	–	–	+/–	–	+-
Technical details						
Performance outdoors/indoors	+/–	+/–	+/–	+/–	–	+
Charge capacity	–	+	–	+	–	+
Efficiency	+/–	+/–	+/–	+	–	+
Adjustability	–	–	–	+/–	–	+
Durability	–	+	+	–	–	+
Sustainability	+	+	+	+	+/–	+
Environmentally friendly character	+	+	+	+	+	+

6.5 DISCUSSION AND CONCLUSIONS

The previously described study (Apostolou 2016) focused on lead-users' interactions with PV products through a practice-oriented approach. A questionnaire was used to identify users' needs and expectations from the PV products and the methods of self- and direct observation for the investigation of user behavior during the interaction. This study was a quite difficult and challenging task, and the combination of various methods was necessary for reliable results. Therefore, in this study, authors conducted not only field trials, but also technical tests for a better understanding of the PV technology by the users.

The tested sample of users for the observation of their behavior with the PV products consisted of 100 students of Industrial Design Engineering Department of Technical University of Delft. The specific sample used quite high standards for the characterization of the products' qualities and offered a critical view of the products' usabilities, designs, and performances. It seems that the tested sample of lead users had a greater critical look than a regular user, due to their educational background in the field of product design and being more ahead of other students with less-relevant educational experience. The specific user type of this study cannot be represented as a regular user or consumer. This user may be considered as a "lead user," since he/she was asked to follow some specific tasks for the evaluation of the products, which might not be recognizable by a regular user. Moreover, the "lead users" of this study proposed solutions and ideas about redesigning the PV products, which is pretty uncommon for regular users to provide such a feedback. On the one hand, lead users can notice and forecast problems that might occur in the future, but on the other hand, due to their educational background and their knowledge in the field of product design and engineering, they understand the boundaries of design and technology in the products. These features are not visible and easily understandable by regular users, who usually criticize the outlook, usability, and performance of the products without caring about the above-mentioned limits. Hence, the beliefs of the lead users in this study do not reflect the real behavior of a simple user, but they could be quite influential regarding the future successful use of the PV products.

The results revealed that the usability, design, aesthetics, and performance of a PV product are quite important factors for lead users. Respondents are quite enthusiastic about PV products if useful and functional, but they need more reliable PV products with a more appealing design.

It was noticed that lead users' expectations before use and their experience afterward deviated significantly. Quantitatively, results show that approximately 40% of the respondents were disappointed with the PV product that they used, 38% found the product useless, 60% believed that the design of the product was of low quality, 88% of the respondents would not buy the PV product, and 70% believed that the price of the PV product did not match with its quality and performance. It is remarkable to notice that approximately 66% of the respondents would prefer a product that can be charged by a cable with a plug, rather than a PV-powered product.

Going back to the theory by Sanders (2002) about the different user-experience characteristics, we tried to distinguish the four categories of the "obvious," the "visible," the "unspoken," and the "dormant" features of users. On the one hand, observing the lead users interacting with the products easily identified the "obvious" and "visible" features. First, the "unspoken" and the "dormant" were investigated through questions regarding users' thoughts before, during, and after the field trial. Lead users enjoyed the benefit to actively interact with the products and criticize products' characteristics, such as the design, usability, performance, aesthetics, or any other feature that was important for them. Furthermore, it was interesting to notice what lead users believe regarding the significance of these products and what they propose for a possible product's redesign. Last but not least, in this study we uncovered lead users' behaviors while interacting with PV products, and we focused mainly on the "unspoken" and "dormant" features, which are important for designers.

The testing sample is limited (six PV products) and general conclusions cannot be drawn. However, results are important because they represent part of the PV products, which are commercially available and easily accessible to consumers and basic user behavior with them. Since the survey outcomes are strongly affected by the type of the specific user, it is not approved that regular users will have similar behavior to the product's use. Therefore, the specific results could not be extended to all target groups. To sum up, the impressions of the lead users about the PV products are not necessarily analogous to the regular users (Apostolou 2016; Apostolou and Reinders 2016). Nevertheless, the results of this study and the specific users' reflections could inspire the future design and usability of PV products. It is believed that the findings of this study will be valuable for designers toward a better understanding of the user behavior, and combined with technical data of PV products, the findings could be used for the design of high-efficient PV products (Apostolou 2016; Apostolou and Reinders 2016).

REFERENCES

Apostolou, G., Design Features of Product-Integrated PV: An Evaluation of Various Factors under Indoor Irradiance Conditions, PhD Thesis Industrial Design Engineering Department, Technical University of Delft (TU Delft), Delft, the Netherlands, 2016.

Apostolou, G. and Reinders, A.H.M.E. (2016) How do users interact with photovoltaic-powered products? Investigating 100 "lead-users" and 6 PV products, *Journal of Design Research*, Vol. 14, No. 1, pp. 66–93.

Apostolou, G. and Reinders, A.H.M.E. (2012) A Comparison of Design Features of 80 PV-Powered Products, *Proceedings of the 27th EUPVSEC, Poster Presentation*, Frankfurt 2012, Germany, pp. 4227–4232.

Apostolou, G., Reinders, A.H.M.E. (2015) Users' Interaction with PV-powered Products: An Evaluation of 6 Products by 100 End-users, *Proceeding of 42nd IEEE Photovoltaic Specialists Conference*, June 14–19, 2015, New Orleans, LA.

Apostolou. G., Reinders, A.H.M.E. (2014) Overview of design issues in product-integrated photovoltaics, *Energy Technology*. Vol. 2, pp. 229–242. doi:10.1002/ente.201300158.

Apostolou, G., Reinders, A., Geurs, K. (2018a) An overview of existing experiences with solar-powered e-bikes. *Energies* 2018, 11(8), 2129.

Apostolou, G., Geurs, K.T., Reinders, A.H.M.E. (2018b) Technical performance and user aspects of solar powered e-bikes—Results of a field study in The Netherlands, University of Twente, Enschede, the Netherlands, *Oral Presentation at ESEIA Conference*, Dublin Energy Lab, April 10–12, 2018, Dublin, Ireland.

Apostolou, G., Verwaal, M. and Reinders, A.H.M.E. (2014) Modeling the performance of product integrated photovoltaic (PIPV) cells indoors, *Proceedings of 26th European Photovoltaic Solar Energy Conference (EU PVSEC)*, Amsterdam, the Netherlands, pp. 3535–3540.

Bakker, C.A., van Eijk, D.J., Silvester, S., Reitenbach, M.E.F.R., de Jong, A.M., Keyson, D.V. and Scott, K. (2010) Understanding and modelling user behaviour in relation to sustainable innovations: The living lab method, in Horvath, I., Mandorli, F. and Rusak, Z. (Eds.): *Proceedings of TMCE 2010 Symposium*, pp. 1–12.

Beurer health and well-being, Beurer kitchen weight scales, https://www.beurer.com/web/gb/products/wellbeing/weight-and-diagnosis/kitchen-scale/, last accessed on October 23, 2019.

Fulton, S.J. (2003) The experience evolution: Developments in design practice, *The Design Journal*, Vol. 6, No. 2, pp. 39–48.

Inhabitat, Product review: Inhabitat Tries Out Ikea's Solar-Powered Sunnan Lamp, https://inhabitat.com/product-review-inhabitat-tries-out-ikeas-solar-powered-sunnan-lamp/ikea-sunnan-review-2/, last accessed on October 23, 2019.

Jacksohn, A., Grösche, P., Rehdanz, K. and Schröder, C. (2019) Drivers of renewable technology adoption in the household sector, *Energy Economics*, Vol. 81, June 2019, Pages 216–226.

Jelsma, J. (2006) Designing 'moralized' products; theory and practice, in Verbeek, P.P. and Slob, A. (Eds.): *User Behavior and Technology Development; Shaping Sustainable Relations between Consumers and Technologies*, pp. 221–231, Springer, Dordrecht, the Netherlands.

Jelsma, J. and Knot, M. (2002) Designing environmentally efficient services; a "script" approach, *The Journal of Sustainable Product Design*, Vol. 2, No. 3, pp. 119–130.

Kakee, S. (2008) Co-designing sustainable user practices; An Exploratory Study for the Living Lab Project, MSc Thesis, Industrial Ecology, Delft University of Technology, Delft, the Netherlands.

Kurdgelashvili, L., Shih, C.-H., Yang, F. and Garg, M. (2019) An empirical analysis of county-level residential PV adoption in California, *Technological Forecasting and Social Change*, Vol. 139, February 2019, Pages 321–333.

Little Sun, Little Sun Original Solar Lamp, https://littlesun.com/product/little-sun-original/, last accessed on October 23, 2019.

Logitech, Wireless Solar Keyboard K750, https://www.logitech.com/en-roeu/product/k750-keyboard-for-business, last accessed on October 23, 2019.

Reich, N.H. et al. (2007) A solar powered wireless computer mouse: Design, assembly and preliminary testing of 15 prototypes, *Proceedings of 22st European Photovoltaic Solar Energy Conference*, Milan 2007, Italy.

Reich, N.H., Elzen, B., Netten, M.P., van Sark, W.G.J.H.M., Alsema, E.A., Veefkind, M. and Silvester, S. (2008) Practical experiences with the PV powered computer mouse "sole-mio," Components for PV Systems, Engineering and System Integration in *Proceedings of 23rd European Photovoltaic Solar Energy Conference*, 2008, Valencia, Spain, pp. 3121–3125.

Reinders, A.H.M.E. and Apostolou, G. (2017) Product Integrated PV, Book Chapter in *Photovoltaic Solar Energy: From Fundamentals to Applications*, Eds. Reinders, A.H.M.E., Verlinden, P., van Sark, W.G.J.H.M., Freundlich, A., John Wiley & Sons, London, UK, January 2017.

Reinders, A.H.M.E. and van Sark, W.G.J.H.M. (2012a) Product-integrated photovoltaics, in Sayigh, A. (Ed.): *Comprehensive Renewable Energy*, Vol. 1, pp. 709–732, Elsevier, Oxford.

Reinders, A.H.M.E., Diehl, J.C. and Brezet, H. (2012b) *The Power of Design: Product Innovation in Sustainable Energy Technologies*, John Wiley & Sons, Delft, the Netherlands, October 2012.

Rodríguez, E. and Boks, C. (2005) How design of products affects user behaviour and vice versa: The environmental implications, *Proceedings of EcoDesign 2005*, Union of Ecodesigners, Tokyo, Japan, pp. 1–8.

Rooden, T. and Kanis, H. (2000) Anticipation of usability problems by practitioners, *Proceedings of the Human Factors and Ergonomics Society 44th Annual Meeting*, San Diego, CA.

Sanders, E.B.N. (2002) From user-centred to participatory design approaches, in Frascara, J. (Eds.): *Design and the Social Sciences*, Taylor & Francis Group, London, UK.

Sanders, E.B.N. (2006a) Design Serving People, Copenhagen Cumulus Working Papers, University of Art and Design Helsinki.

Sanders, E.B.N. (2006b) Design research in 2006, *Design Research Quarterly*, Vol. 1, No. 1, pp.1–8.

Shackel, B. (1984) The concept of usability, in Bennet, J. et al. (Eds.): *Visual Display Terminals*, pp. 45–88, Prentice-Hall, Englewood Cliffs, NJ.

Smit, E., Stevels, A. and Sherwin, C. (2002) User centred ecodesign: Experiences from the design of a human-powered remote control, *Proceedings of Care Innovation 2002*, November 25–28, IEEE, Vienna.

van Mierlo, B. (2017) People's involvement in residential PV and their experiences, in Reinders, A., Verlinden, P., van Sark, W., Freundlich, A. (Eds.): *Photovoltaic Solar Energy: From Fundamentals to Applications*, 1st edn, John Wiley & Sons, New York.

von Hippel, E. (1982) Appropriability of innovation benefit as a predictor of the source of innovation, *Research Policy*, Vol. 11, No. 2, pp. 95–115.

von Hippel, E. (1986) Lead users: A source of novel product concepts, *Management Science*, Vol. 32, No. 7, pp. 791–805.

Vasseur, V. and Kemp, R. (2015) The adoption of PV in the Netherlands: A statistical analysis of adoption factors, *Renewable and Sustainable Energy Reviews*, Vol. 41, Pages 483–494.

Vredenburg, K., Isensee, S. et al. (2002) *User-Centered Design; An Integrated Approach*, Prentice-Hall, Upper Saddle River, NJ.

Waka Waka Go Solar, Waka Waka light and Waka Waka Power, https://waka-waka.com/, last accessed on October 23, 2019.

Wever, R., van Kuijk, J. and Boks, C. (2008) User-centred design for sustainable behaviour, *International Journal of Sustainable Engineering*, Vol. 1, No. 1, pp. 9–20.

Design Case 6

PV-Powered Charging Station for E-Bikes

Cihan Gerçek

6.1 INTRODUCTION

This chapter presents a design case of a photovoltaic (PV)-powered charging station for electric bikes(e-bikes), which has been realized in 2018 at the University of Twente (UT) in the framework of Living Smart Campus projects. Living Smart Campus enables researchers to test, demonstrate, and realize their innovations within the campus while improving campus life. An example of such an innovation is a PV charging station for e-bikes, to enable charging of e-bikes in a more environmentally friendly way than charging by the grid. This design case of the PV-powered charging station and its features is an example of multi-criteria decision making in design processes for solar-powered products, taking into account the load for e-bike charging, solar-energy generation, cost-effectiveness, self-consumption, and user aspects. Figure DC-6.1 presents the solar charging station and the solar e-bikes used for the tests of the performance of this PV charging station.

Solar-powered electric mobility can contribute to reduced greenhouse-gas emissions (Bhatti et al. 2016). In this framework, it is interesting to explore how e-bikes' batteries can be charged by PV power. There exist two main ways to realize solar-powered e-bikes:

- Solar e-bikes: In this bike, PV solar cells are integrated in its parts, charging its batteries with solar energy.
- PV-powered charging station: PV modules or PV cells are integrated in an electric charging station for e-bikes, which transforms the solar energy into electricity to be fed into the chargers of the parked e-bikes.

(a) (b)

FIGURE DC-6.1 (a) Solar e-bike, and (b) PV-powered charging station in UT campus.

In previous studies, different aspects of solar-powered e-bikes were evaluated (Zhang et al. 2019; Apostolou et al. 2018). A field experiment was conducted with solar e-bikes (Figure DC-6.1) of the brand Sparta M7S LTD with six flexible PV mini-modules with 18 copper indium gallium diselenide (CIGS) cells each that were integrated in the front wheel (Estimated nominal PV power: 66–72 Wp at Standard Test Conditions) (Apostolou et al. 2018). During cloudy weather, these integrated PV modules were estimated to produce approximately 1 to 10 W depending on the time and the wheel orientation, which is quite low (Apostolou et al. 2018). Another constraint is the aerodynamics and wind stability, as the drag surface of the front wheel is increased by PV cells, which might deteriorate the aerodynamics of the bike and be problematic while cycling against the wind (Apostolou et al. 2018). The study also pointed out that commercially available solar e-bikes are rare, and their prices are substantially higher than conventional bikes and higher than e-bikes. Considering the low power input and additional cost investments, it could be concluded that PV charging stations would be more appropriate for charging of e-bikes than integrated solar PV modules.

6.2 DESIGN OF A PV-POWERED CHARGING STATION

Regardless of the situation, a well-designed charging station delivers in a continuous and reliable manner, the required electricity to the chargers to convey energy to the storage of the subject device, in this case, e-bikes. If charging stations of e-bikes succeed to meet technical requirements even in worst scenarios and are aesthetically appealing and well located, they will encourage users to electric mobility. PV-powered charging station will increase the share of sustainable energy in electric mobility; however, the solar energy output is highly dependent on atmospheric conditions. To avoid any interruption of the charging process, storage or additional power supply would be required. With a connection to the grid, the station will continuously supply the charging regardless of the time and the atmospheric conditions. Nevertheless, PV systems of the station should be optimally designed and accustomed to the demand, so as less energy as possible needs to be imported, as a consequence, the share of self-sufficiency and self-consumption stays relatively high.

First, we will specify the definition of e-bikes according to EU regulations, following the design features of charging station considering location and users aspects will be defined. Second, a specific flow chart would summarize the technical design steps. After the electrical demand/charging load scenario is defined, the choice of the PV system will be detailed and simulated for a typical year, month, and day. Last but not least, the demand-supply match will be modeled in order to highlight the self-sufficiency, self-consumption, and system losses. We will conclude with the final specifications and features, cost estimations, and payback time of investments, and some perspective.

6.2.1 Definition of E-Bikes according to EU Regulations

There is a big variety of e-bikes, and the diverseness might even create confusions to distinguish them from e-scooters. In this text, e-bike refers to electrically powered assisted cycle (EPAC), also called pedalec, in which the motor is only activated while the user is pedaling. Speed and other characteristic limits vary based on the legislation and regulation of the country. EU regulations limits the power of the electric motor to 0.25 kW and speed limit to 25 km/h in order to be classified as a bicycle (EU Parliament and EU Council 2013). In the Netherlands, the eligibility criteria are same to bike through the dedicated bike lanes. Further characteristics and classifications might be found in the same regulation (EU Parliament and EU Council 2013) or elsewhere (European Cyclists' Federation; Boedijn and Kremer 2019; Statistics Netherlands [CBS] 2016a).

6.2.2 Choice of the Location within the Campus Considering User Aspects

The UT campus has a wide area, which is estimated around 146 hectares. The best location for the installation of a PV charging station would be a south-orienting one, without shadings, meaning without high buildings and trees in the surroundings. The Spiegel Building in Figure DC-6.2 is mainly used for administration purposes; it is a 6th-floor building (University of Twente 2019) with approximately 25 m height (Figure DC-6.2). It has one of the biggest bicycle parking spots, which is regularly used, next to one of the most frequent bike lanes and urban roads in the region, connecting Hengelo to Enschede (Hengelosestraat—Figure DC-6.2). A part of the old bicycle parking spot is used for the PV-powered charging station in order to avoid any major modification over the area. The aim in such a location choice is visibility, easy access to all citizens, maximizing the output by a south-facing orientation, and avoidance of shading coming from the same orientation.

6.2.3 Number of E-Bike Charging Slots

Neither number of e-bikes parked annually nor the statistical data about the number of people using the buildings are known. General information and national average values would be used in order to estimate the number of e-bikes

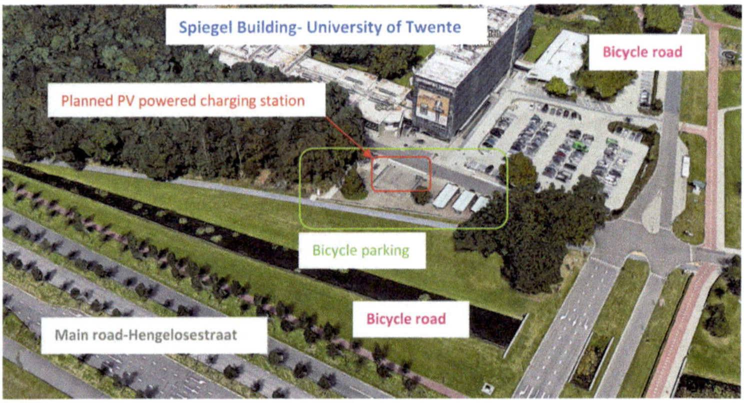

FIGURE DC-6.2 Image of the planned location using Google Earth before the construction of the charging station occurred. (Google Earth, University of Twente- Spiegel Building;52.23909884,6.84929937, Available from: https://earth.google.com/web/@52.23909884,6.84929937,25.25039077a,249.59537236d,35y,0.00000001h,44.99766433t,-0r.)

potentially parked next to the buildings in order to define the potential need of the number of charging slots on the PV-powered charging station.

In Equation 6.1, the formula of potential number of e-bikes parked per building, N, for an average day is given:

$$N = n \times b \times p \tag{6.1}$$

whereas n is number people per building, b is the share of bikes in commuting, and p is the share of e-bikes over the population. Considering the number of people enrolled or working (13,509, excluding visitors) (University of Twente 2019) divided by the number of enumerated buildings at UT, 65 as indicated at map in Figure DC-6.3 (University of Twente 2019), the number of people per building might be roughly estimated as 207 people. Dutch average of preference to commute by bike is 25%, calculated by Netherlands Institute for Transport Policy Analysis (KiM) (2018) based on a national survey in 2016 (Statistics Netherlands [CBS] 2016b). The Dutch share of e-bikes over the population has been calculated 12.5% in 2016 (Ministry of Infrastructure and Water management 2018). Using Equation 1, the number of people expected for the building and characteristics for the Netherlands, only six e-bike trips (round) would be potentially made for each building daily for UT.

6.2.4 Flowchart

Figure DC-6.4 shows the flowchart employed in order to evaluate the energy performance of the PV charging station, which has been installed at UT's campus. The method has been adapted from an EV charging station study (Good et al. 2019), based on IEA Task 7 recommendations (International Energy Agency 2002). The power production of the PV system and the power consumption due to an e-bike's battery charging is represented in two different branches, and following sections will detail these steps.

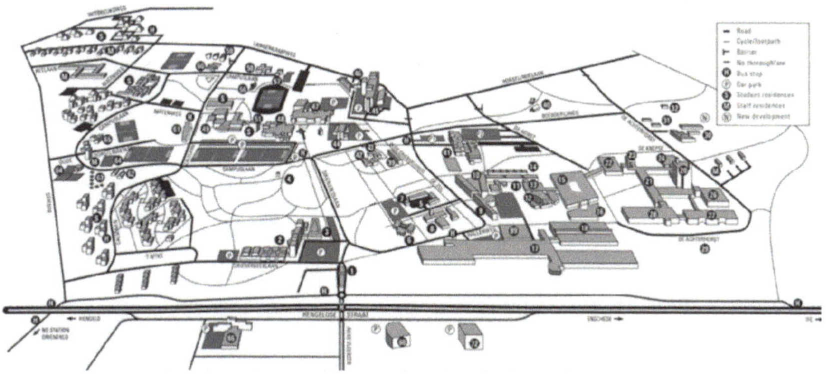

FIGURE DC-6.3 Plan of the University of Twente campus over 146 hectares.

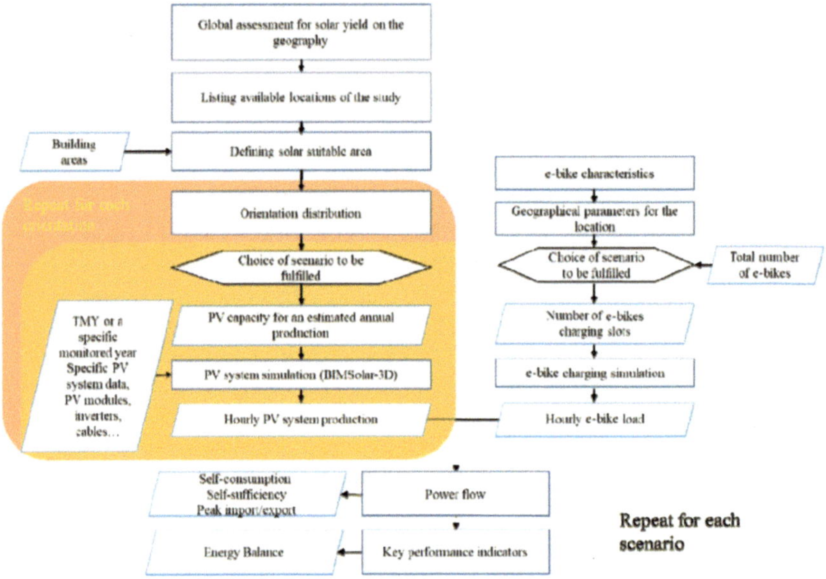

FIGURE DC-6.4 Flowchart summarizing different steps for the technical design. (Adapted from Good, C. et al., *Energy*, 168, 111–125, 2019.)

6.2.5 Characteristics of E-Bikes from an Energy Perspective

Based on the commercially available bikes and literature (Apostolou et al. 2018; Bosch 2019a, 2019b), Table DC-1.1 resumes the main characteristics of commercially available e-bikes (EPAC).

Field tests have been effectuated with five Sparta M7S LTD e-bikes, with integrated PV, battery capacity of 400 Wh (36V 11 Ah), and a charging time of 2.5 hours with charging rate of 160 W. With the aging of the batteries, the field

TABLE DC-1.1
Main Characteristics of E-Bikes (EPAC)

Parameter	Range	Average
Motor and battery (V)	24–52	36
Battery capacity (Ah)	3–23	11
Battery capacity (Wh)	100–1200	400
Charging current (A)	0.2 up to 4	2.5
Nominal charging (W)	100–260	160
Charging time from empty to full	45 min. up to 5.3 hours	2.5–3 hours
Partial charging (empty to +85% full)[a]	30 min up to 4.5 hours	2 hours

[a] Time spent in bulk charging phase until it switches to constant voltage charging (which is slower charging phase): 0% state of charge up to ~85%–95% state of charge of the e-bike battery.

test and the charging currents showed that it drops for some bikes as low as 90 W because the batteries are already 5 years old. Thus, 160 W would be taken as the scenario, but in practice, this would be lower.

6.2.6 Amount of Energy and Time Required for Charging

The charging scenario and time highly depend on the e-bike charger current and the capacity of the battery, state of charge (SOC), and therefore the range of the trip. The field-test survey indicated approximately 4% of 327 trips studied were 55–56 km distance, 96% was below 20 km, and the average distance was 10.3 km (Apostolou et al. 2018). According to an online simulation tool for e-bike range assistant (Bosch 2019a), ranges more than 50 km could be reached with 400 Wh, which gives the worst-case amount of energy to be charged daily per e-bike.

Possible worst charging scenario in time would be that each charging slot would be used twice a day for charging purposes, morning and afternoon. In that scenario, the average situation over the charging slots would be approximately 160 W required and 2 hours of charging until 85% of SOC. Charging of the 85% to 100% SOC will be neglected because the required power will diminish by time.

Momentary worst scenario would be six e-bikes charging at the same time; 960W instantaneous power would be needed. Each charging slot being used for 2 hours twice (morning and afternoon) would result in 1920 Wh energy charged per day in total. Working days usage (excluding 5 weeks of leave and national holidays) would result in 434 kWh annual energy.

6.3 ANNUAL, MONTHLY, AND DAILY SOLAR YIELD ANALYSIS FOR THE LOCATION

Solar geographical irradiance of the exact spot is investigated using PVGIS, an interactive tool of the EU Commission. The optimized azimuth angle is −6°, and the tilt angle 38°. Yearly "plane of array" irradiation is 1230 kWh/m^2 for the typical meteorological year (TMY) (European Commission 2017). With azimuth angle of −30° of the building, the value slightly decreases to 1210 kWh/m^2. However, solar irradiance should be assessed in 3D in order to see the impacts of the possible far shading objects, heat losses, wire losses, and other detailed information that PVGIS cannot provide. Energy demand of charging should be assessed and demand–supply should be optimized to increase self-consumption. In Figure DC-6.5, the red arc represents the solar path during January. For such a low-degree solar path, any height results in yield losses, including the building height, which highlights the importance of modeling in 3D. For shading purposes, a panoramic photo is taken (see Figure DC-6.6), facing the south from the exact location of the PV-powered charging station, in order to indicate the surrounding objects and trees. Respecting their location, the trees significant to cause shading are added in 3D geometry, respecting their true dimensions, diameters, and distances (Google Earth). Next, the expected charging-station dimensions are added in 3D for the analysis of the irradiance on the solar charging station. In that stage, the charging station will be represented as a rectangular box with dimensions 6.0 m × 2 m × 2 m. Since a typical bicycle length is 1.7 m, the space between the parking bars is approximately 1 m, and six charging slots

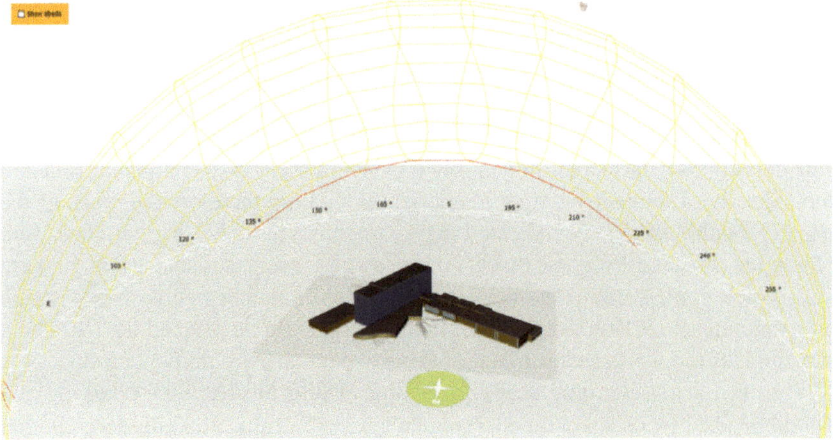

FIGURE DC-6.5 Spiegel Building and Solar Path for a typical meteorological year obtained by importing PVGIS satellite data for the exact location and 3D model of the building in software BIM Solar.

FIGURE DC-6.6 South view of the PV-powered charging station planned location, which is bike parking next to the Spiegel Building-UT, where shading objects are visible, mainly, trees at the right-side and left-side of the panoramic photo.

were assumed with 38° optimized slope canopy, which allows optimum PV performance. Figure DC-6.7 is the yearly irradiance simulation results for a TMY. Despite all the shading objects included, the annual yield is close to 1000 kWh/m^2 over the PV-powered charging station spot. This is 20% less than what has been calculated by PVGIS because the shadows were not taken into account. The solar path of the location in Figure DC-6.5 varies from 15° to 61°, from December to June, which explains this significant difference in the results due to shadow.

The daily average irradiance is 220 Wh/m^2. We mentioned previously that the scenario was nominal charging of 160 W and maximal 260 W, which requires 960–1600 W instantaneous power. The annual energy requirement was 434 kWh, and close to 2 kWh daily. Considering average daily irradiance (220 Wh/m^2), less than ~10 m^2 of surface for PV system is needed for the charging station.

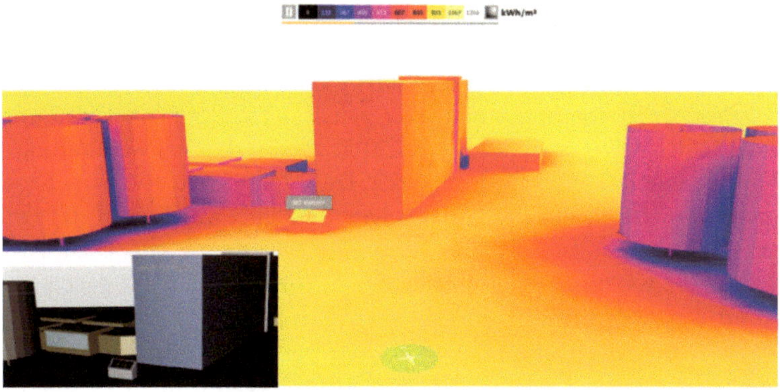

FIGURE DC-6.7 Spiegel Building and trees in surroundings in 3D, and the solar yield in kWh/m^2 for a typical meteorological year.

6.4 ELECTRICAL CIRCUIT OF THE PV SYSTEM

The electrical circuit schematic of a typical grid-connected PV charging system is shown in Figure DC-6.8. It contains PV panels, connected to an inverter, which effectuates at DC side maximum power point tracking (MPPT) in order to maximize the PV generation output, and transforms it to AC, so that the energy can be used for charging purposes, and the surplus of the energy will be sent to the grid. It is also possible to develop a charging system with a battery to be independent of the grid. This solution requires a converter with MPPT, a battery charger, a battery, and an inverter to supply AC power so that the e-bike chargers can be plugged in and charged with solar power. The system might operate even without any grid supply depending on the battery and PV size. However, this configuration will result in higher energy losses, higher levelized cost of electricity (Kost et al. 2018), and higher greenhouse-gas emissions as compared to grid-connected systems.

6.4.1 Specification of the PV Modules, Inverter, and Cabling

An extensive database called Photon Database, comprehending more than 50,000 PV, and approximately 7,500 inverters dedicated for PV has been consulted to see the good match (Photon Laboratory 2019). Because Dutch climate

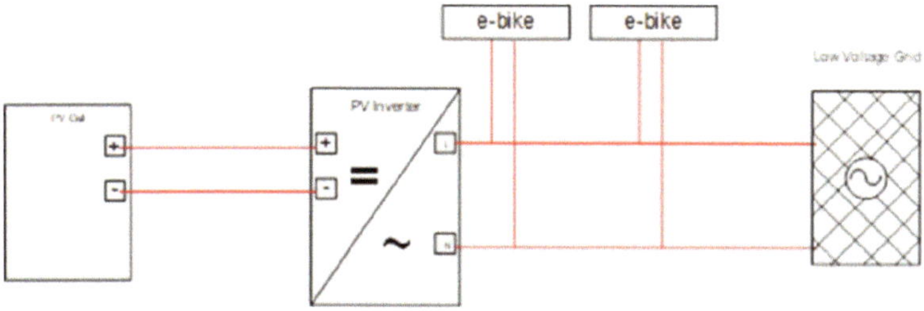

FIGURE DC-6.8 Circuit schematic of a grid-connected PV charging station.

and irradiation do not have a specific requirement against losses, a cost-effective solution would be choosing crystalline silicon technology. Both the maximum nominal charging power of e-bikes (260W) and daily average irradiance have been considered for the choice of the PV panels. Hence, six panels of Canadian Solar ALL-BLACK CS6K-270, 275, 280M, 60 dark mono-crystalline cells 275Wp with a module efficiency of 17.11%. Dim. 1650 × 992 × 40 mm anodized aluminum alloy frame are chosen. They have a product warranty of 10 years and a linear power output warranty of 25 years. The exact same panels are virtually inserted on the top of the charging station in BIMsolar. The six panels cover a 9.8-m^2 surface, and software estimates an annual production of 1487 kWh.

The chosen PV inverter is X1Series SOLAX Power, 1.1 kW–2.0 kW. It converts the DC current of the PV generator (Euro eff. 96%) with MPPT (eff. 99.9%) into AC current and feeds it into the public grid with total harmonic distortion of 2%. For power quality, resonance, harmonic distortion, and other issues, readers can refer to our previous publication (Zhang et al. 2019) and datasheet of the product (Solax Power 2019). This specific inverter was not in the photon database so an equivalent one is used, with a 2-m cabling distance from the first PV module to inverter. All PV panels are connected in series, which means only one MPPT is done by the inverter, as it is the case in reality.

6.5 SIMULATION RESULTS

6.5.1 Simulation Results for the PV System

The PV system DC production and diverse losses are given for monthly basis in Figure DC-6.9. The most consequent loss is the 8.8% heat loss, 5.7% shadow loss, and mismatching, and cable losses are negligible (less than 1%). Production for an average day of each month is given in Figure DC-6.10.

6.5.2 Simulation Results for Self-Consumption and Cost Analysis

The Dutch commuting by working persons curve has two spikes occurring slightly before 9:00 and 17:00 (Statistics Netherlands [CBS] 2016a). Assuming

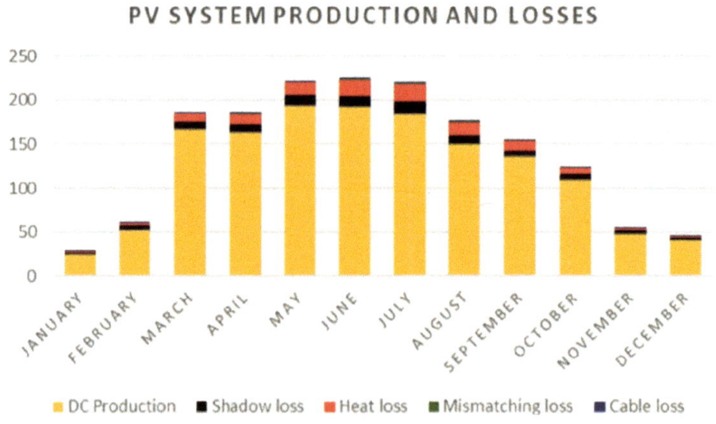

FIGURE DC-6.9 PV system monthly production and losses for TMY.

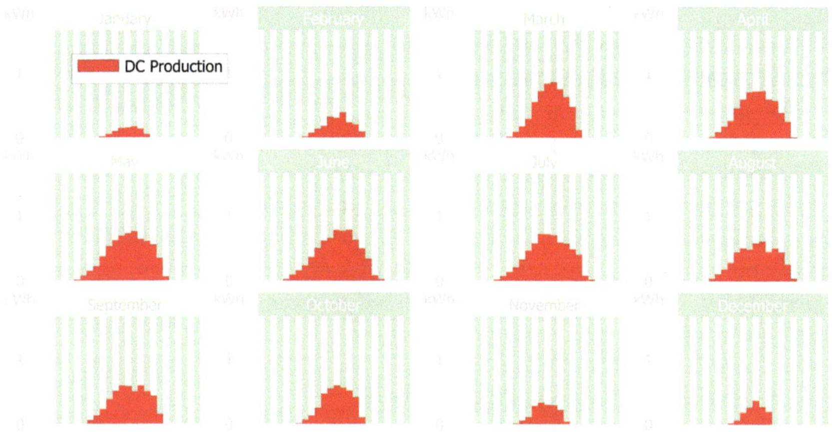

FIGURE DC-6.10 PV system generation for the average day of the month at hourly resolution for TMY.

434 kWh annual charging energy with a homogeneous charging pattern only occurring between 9:00 to 17:00, it would lead to approximately 78% of annual self-sufficiency, meaning the solar energy share of the charged e-bikes is relatively high. The price of the electricity is taken 15 c€/kWh, with net metering. Including maintenance cost, according to BIMsolar, the PV charging station will pay itself back in 12 years with a 66% of investment ratio in 20 years.

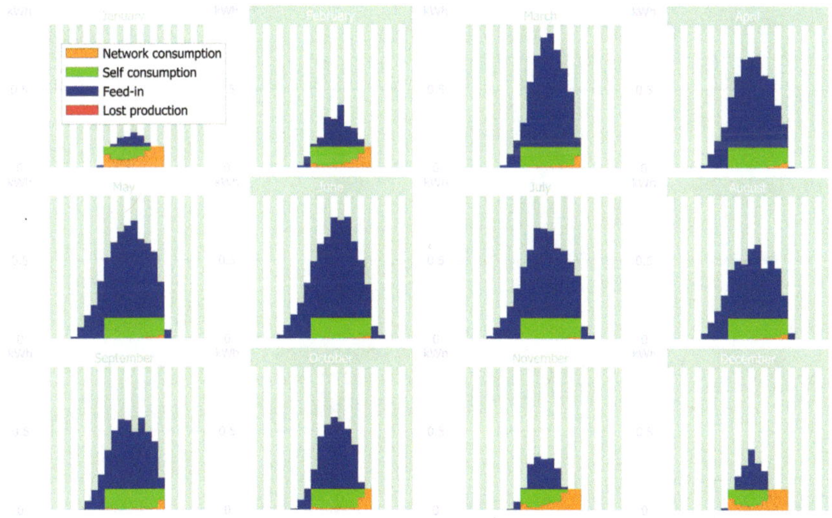

FIGURE DC-6.11 PV-powered charging station monthly averages with constant charging between 9:00 and 17:00: The import from the grid (network consumption: orange), amount of solar energy used for charging e-bikes (self-consumption: green), and amount of solar energy sent to the building or to the grid (feed-in: blue).

Figure DC-6.11 shows the results of the charging station (excluding building-consumption patterns), feed-in (blue), self-consumption (green), and the energy that had to be supplied/imported from the grid (network—orange).

The self-consumption is 23%, and the surplus of energy is shown as feed-in. As connected to the Spiegel Building, the building will consume the most of the feed-in energy. The estimated interior surface of the Spiegel Building is 12000 m^2, and PV charging station is linked to the building via a low-voltage transformer located to 50 m away. The average consumption of the Dutch university buildings vary between 31 and 141 $kWh/m^2/year$, and even the minimal consumption 31 $kWh/m^2/year$ taken, the self-consumption rises to 90%, independently from the e-bikes charging activities.

6.6 CONCLUSIONS AND PERSPECTIVE

Respecting the users aspects and technical requirements summarized in the flowchart (Figure DC-6.4), step by step, we investigated how to use statistical data in order to give an average value of consumption and design a PV system accustomed to the aimed consumption. The PV charging station is investigated in its 3D environment, with the exact same PV technology, an equivalent inverter, and wiring configuration. The shading, wiring, mismatch, and heat losses are taken into account for the end results. We have been able to estimate the percentage of the self-consumption of the PV charging station with the given conditions of TMY and scenarios. The charging station in its operational installation

FIGURE DC-6.12 PV Charging station next to the Spiegel Building.

is shown in Figure DC-6.12. Aesthetical aspects and other functional aspects have been taken into account, the sunflower icon and the futuristic design shows the sustainability aims of the charging station and respect of the surrounding architecture.

One of the functional problems in an open space e-bike charging system was the fact that the chargers were open to thievery and vandalism; installation of dedicated boxes or chains for chargers with borrowable keys might be a solution. One might also measure accurately the demand profile and/or power flow, or use more extended survey results to estimate when people will park their bikes for charging to obtain the exact load profiles before designing a PV system. It goes without saying that low voltage transformer of the building that charging stations is coupled with, should have the required power capacity to grasp the additional load.

REFERENCES

Apostolou, G., Reinders, A. and Geurs, K. (2018). An overview of existing experiences with solar-powered e-bikes. *Energies* Vol. 11, Pages 2129.

Bhatti, A.R., Salam, Z., Aziz, M.J.B.A. and Yee, K.P. (2016). A critical review of electric vehicle charging using solar photovoltaic: A critical review of PV-EV charging. *International Journal of Energy Research* Vol. 40, Pages 439–461.

Boedijn, S. and Kremer, M. (2019). Rijwiel en Automobiel Industrie (RAI) Association-Bicycles. Available from: https://raivereniging.nl/fietsen.

Bosch. (2019a). E-bike Range Assistant. Available from: https://www.bosch-ebike.com/en/service/range-assistant/.

Bosch. (2019b). How long does it take to charge the battery? Available from: https://www.bosch-ebike.com/en/service/faq/how-long-does-it-take-to-charge-the-battery/.

EU Parliament and EU Council. (2013). Regulation No 168/2013 on the approval and market surveillance of two- or three-wheel vehicles and quadricycles. European Parliament and Council, Brussels, Belgium. Available from: https://eur-lex.europa.eu/legal-content/EN/ALL/?uri=CELEX%3A32002L0024.

European Commission. (2017). Photovoltaic geographical information system. EU Commission-PVGIS. Available from: http://re.jrc.ec.europa.eu/pvg_tools/en/tools.html#MR.

European Cyclists' Federation. Electric-Bicycle (pedelec) Regulation. Available from: https://ecf.com/what-we-do/road-safety/electric-bicycle-pedelec-regulation.

Good, C., Shepero, M., Munkhammar, J. and Boström, T. (2019). Scenario-based modelling of the potential for solar energy charging of electric vehicles in two Scandinavian cities. *Energy* Vol. 168, Pages 111–125.

Google Earth. University of Twente- Spiegel Building;52.23909884,6.84929937. Available from: https://earth.google.com/web/@52.23909884,6.84929937,25.25039077a,249.59537236d,35y,0.00000001h,44.99766433t,-0r.

InternationalEnergyAgency.(2002).PhotovoltaicPowerSystemsProgrammeT7-4:Potential for building integrated photovoltaics. IEA, Ursen, Switzerland. Available from: http://www.iea-pvps.org/index.php?id=9&eID=dam_frontend_push&docID=394.

Kost, C., Shammugam, S., Jülch, V., Nguyen, H.-T. and Schlegl, T. (2018). Levelized Cost of Electricity Renewable Energy Technologies. Fraunhofer Institute for Solar Energy Systems, Freiburg, Germany.

Ministry of Infrastructure and Water management. (2018). User characteristics and trip patterns of e-bike use in the Netherlands Results from the Dutch National Travel Survey and the Mobility Panel Netherlands. Available from: https://www.kimnet.

nl/binaries/kimnet/documenten/presentaties/2018/04/10/user-characteristics-and-trip-patterns-of-e-bike-use-in-the-netherlands/User+characteristics+and+trip+patterns+of+e-bike+use+in+the+Netherlands.pdf.

Netherlands Institute for Transport Policy Analysis (KiM). (2018). Mobility report. Netherlands Institute for Transport Policy Analysis, The Hague. Available from: https://www.kimnet.nl/binaries/kimnet/documenten/rapporten/2018/10/30/kerncijfers-mobiliteit-2018/kerncijfers+mobiliteit+2018.pdf.

Photon Laboratory. (2019). Photon database. Available from: https://www.photon.info/en/photon-databases.

Solax Power. (2019). X1 series user manual. Available from: https://www.solaxpower.com/wp-content/uploads/2017/01/X1-Mini-Install-Manual.pdf.

Statistics Netherlands (CBS). (2016a). Transport and mobility 2016. Den Haag: Centraal Bureau voor de Statistiek; Available from: https://www.cbs.nl/en-gb/publication/2016/25/transport-and-mobility-2016.

Statistics Netherlands (CBS). (2016b). Netherlands travel survey (OViN). Statistics Netherlands (CBS), The Hague.

University of Twente. (2019). Facts & Figures 2018. Available from: https://www.utwente.nl/en/facts-and-figures/#organization.

Zhang, Z., Gercek, C., Renner, H., Reinders, A., Fickert, L. (2019). Resonance instability of photovoltaic e-bike charging stations: Control parameters analysis, modeling and experiment. *Applied Sciences*. Vol. 9, Pages 252.

Index

Note: Page numbers in italic and bold refer to figures and tables, respectively.

E

F

G

H

I

J

L

M